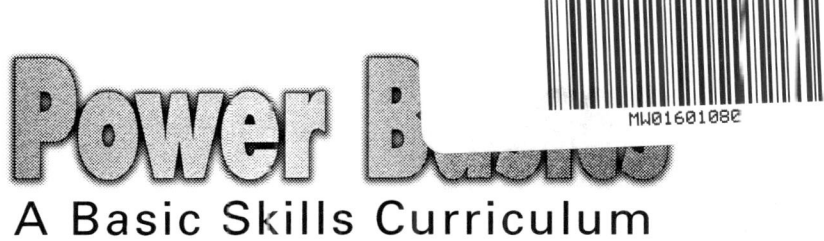

Power Basics

A Basic Skills Curriculum

SCIENCE
Physics Mastery

Included in this volume:
   Forces
   Energy and Heat
   Sound and Light
   Electricity, Magnetism, and Beyond
   Teacher Guide

Robert Taggart

J. WESTON
WALCH
PUBLISHER
Portland, Maine

User's Guide
to
Walch Reproducible Books

As part of our general effort to provide educational materials that are as practical and economical as possible, we have designated this publication a "reproducible book." The designation means that purchase of the book includes purchase of the right to limited reproduction of all pages on which this symbol appears:

Here is the basic Walch policy: We grant to individual purchasers of this book the right to make sufficient copies of reproducible pages for use by all students of a single teacher. This permission is limited to a single teacher, and does not apply to entire schools or school systems, so institutions purchasing the book should pass the permission on to a single teacher. Copying of the book or its parts for resale is prohibited.

Any questions regarding this policy or requests to purchase further reproduction rights should be addressed to:

Permissions Editor
J. Weston Walch, Publisher
321 Valley Street • P. O. Box 658
Portland, Maine 04104-0658

1   2   3   4   5   6   7   8   9   10
ISBN 0-8251-4313-6
Copyright © 1998, 2001
J. Weston Walch, Publisher
P.O. Box 658 • Portland, Maine 04104-0658
www.walch.com
Printed in the United States of America

Table of Contents

Teacher Guide Pages

Reproducible Student Pages

■ **PART 4: ELECTRICITY, MAGNETISM, AND BEYOND** ■

To the Teacher

Welcome to *Physics Mastery*. This is a basic skills course created to take students beyond the basics to a broader understanding of physics. This course was designed for the needs of students working at the secondary level and beyond. This flexible publication includes all of the reproducible student reading and activity pages necessary for a comprehensive review of physics. Also included are teacher guide pages, reproducible extension activities, and reproducible tests to administer on an as-needed basis, as well as answer keys to all student questions.

■ Teacher Guide Pages ■

Teacher Guide Pages at the beginning of this book provide you with strategies for working with a diverse student population. You are first provided with a reproducible diagnostic pretest (Course Mastery Test A) for the entire *Physics Mastery* course. This test will help you pinpoint individual student needs and can help you determine where to start each learner in Walch's *Physics Mastery* curriculum. Pre- and posttests (Mastery Tests A and B) are then provided for **each of the four major parts** in the student pages. These tests allow you to assess student comprehension and content mastery in progressive steps. Finally, Course Mastery Test B is provided as a posttest to assess mastery of the entire *Physics Mastery* course.

The Teacher Guide Pages also include classroom hints and suggested procedures, as well as reproducible extension activities. These extensions are correlated to key sections within the student material. They are designed to help students move beyond the printed page and apply critical lesson content in real-world contexts. Suggestions for additional extension activities (such as Internet applications, research projects, and hands-on activities) are provided to offer maximum flexibility as you adapt this material to meet the diverse needs of your classroom population.

Finally, the Teacher Guide Pages end with a complete answer key to all student practice sections, lesson mastery tests, part mastery tests, and course mastery tests.

■ Reproducible Student Pages ■

Reproducible Student Pages in this book are composed of four major parts:

        Part 1: Forces
        Part 2: Energy and Heat
        Part 3: Sound and Light
        Part 4: Electricity, Magnetism, and Beyond

Whether studied individually (for selected reinforcement) or consecutively (as an entire step-by-step course), the lessons in the student material offer a solid review—with many opportunities for skill practice—of the most important concepts of physics mastery students need to know. In addition, frequent special features like *Tips, Think About It,* and *In Real Life* sections provide your students with extra hints and strategies for mastering the material—and help build critical-thinking skills at the same time.

We are confident that this product from J. Weston Walch, Publisher, will help your students attain the academic success and skills mastery they require to compete and succeed in today's global marketplace.

Teacher
Guide
Pages

■ Physics Mastery
DIAGNOSTIC KEY FOR COURSE MASTERY TESTS

The *Physics Mastery* Course Mastery Tests A and B were designed to be used as diagnostic tools. Once administered, these tests will help you determine which competencies each learner has mastered, enabling you to pinpoint individual learner needs within the *Physics Mastery* course.

The diagnostic matrix below will show you how to analyze an individual learner's course mastery test score to identify the most appropriate *Physics Mastery* section to begin with. *(Note: A learner is determined to have demonstrated competency of a particular section's content when he or she provides correct answers for over 80 percent of the test questions drawn from that section's content.)*

The following matrix applies to both Course Mastery Test A and Course Mastery Test B.

Test questions 1–7:
If learners answer at least **6** of these questions correctly, they have demonstrated mastery of essential content covered in *Part 1, Forces.*

If learners do not answer at least 6 of these questions correctly, they should be directed to the *Introduction to Physics* course in the *Power Basics* Introductory Science series.

Test questions 8–14:
If learners answer at least **6** of these questions correctly, they have demonstrated mastery of essential content covered in *Part 2, Energy and Heat.*

Test questions 15–21:
If learners answer at least **6** of these questions correctly, they have demonstrated mastery of essential content covered in *Part 3, Sound and Light.*

Test questions 22–28:
If learners answer at least **6** of these questions correctly, they have demonstrated mastery of essential content covered in *Part 4, Electricity, Magnetism, and Beyond.*

If learners demonstrate mastery of the essential content contained in all four parts of the *Physics Mastery* course, they should be directed to *Introduction to Life Science*, in the *Power Basics* Science Mastery Series.

■ Physics Mastery
COURSE MASTERY TEST A

Directions: Circle the correct answer to each of the following questions.

1. How can an object change its velocity without changing its speed?

    (a) by balancing changes in acceleration with changes in velocity

    (b) by changing its acceleration

    (c) by changing its direction

    (d) This is impossible.

2. When an object makes one complete rotation, what is its angular displacement?

    (a) 90°

    (b) 180°

    (c) 270°

    (d) 360°

3. What does the principle of inertia state?

    (a) that friction will always overcome velocity

    (b) that the natural position of an object is at rest

    (c) that the natural position of an object is in motion

    (d) that the velocity of an object does not change
    unless a force acts upon it

4. What is the term for an inward perpendicular force that causes an object to move in a circle?

    (a) centrifugal force

    (b) centripetal force

    (c) circular force

    (d) rotary force

5. In a vacuum, which object would fall fastest?

   (a) a feather (0.1 ounce)

   (b) a rubber ball (2 ounces)

   (c) a stone (3 pounds)

   (d) They would fall at the same rate.

6. What determines the gravitational attraction between two objects?

   (a) their density and volume

   (b) their mass and the distance between them

   (c) their volume and the distance between them

   (d) their weight and density

7. Who demonstrated that nothing can travel faster than light?

   (a) Einstein

   (b) Galileo

   (c) Kepler

   (d) Newton

8. Using the term "work" in its scientific sense, which of the following people is not doing any work?

   (a) a boy pulling a wagon down a path

   (b) a girl pushing a doll carriage across a room

   (c) a woman pushing against a wall to hold it in place

   (d) a builder raising a weight with a rope and pulley

9. An arrow with a mass of 0.2 kg is moving at 40 m/sec. How much kinetic energy does it have?

   (a) 40 J

   (b) 80 J

   (c) 120 J

   (d) 160 J

10. What is the term for the situation when heat is no longer being transferred between two objects?

    (a) dynamic equilibrium

    (b) heat balance

    (c) thermal equilibrium

    (d) zero heat exchange

11. When you ice skate, even if the temperature is well below freezing, you are actually skating on a thin layer of water, not ice. Where does this water come from?

    (a) Some water is always present on top of ice, no matter how cold it is.

    (b) The pressure of your skates lowers the melting point of the ice.

    (c) The pressure of your skates raises the melting point of the ice.

    (d) The speed of your skates raises the melting point of the ice.

12. In general, which are the best conductors of heat?

    (a) gases

    (b) liquids

    (c) solids

    (d) All three are about equally effective conductors.

13. What is the term for a device that uses heat to create motion?

    (a) a heat collector

    (b) a heat conductor

    (c) a heat engine

    (d) a heat transfer device

14. What measures the disorder of a system?

    (a) Carnot efficiency

    (b) entropy

    (c) specific heat

    (d) thermal equilibrium

15. What is the term for a vibration that moves from one place to another?

    (a) crest

    (b) peak

    (c) tsunami

    (d) wave

16. What does the speed of a wave depend upon?

    (a) its amplitude

    (b) its frequency

    (c) its medium

    (d) its wavelength

17. Which of the following is a form of electromagnetic radiation?

    (a) radio waves

    (b) visible light

    (c) X-rays

    (d) all of the above

18. Which of the following is a naturally occurring prism or set of prisms?

    (a) lightning

    (b) the northern lights

    (c) a rainbow

    (d) sunspots

19. What is the term for what happens when a ray of light strikes an object and is stopped?

    (a) absorption

    (b) reflection

    (c) refraction

    (d) transmission

20. When light passes into a slower medium, how will it be bent?

    (a) away from the normal

    (b) in the same direction as the normal

    (c) toward the normal

    (d) It will not be bent.

21. What is the term for the production of color by interference?

    (a) diffraction

    (b) ionization

    (c) iridescence

    (d) polarization

22. Which of the following pairs of electrical charges will attract each other?

    (a) negative-negative

    (b) negative-positive

    (c) neutral-neutral

    (d) positive-positive

23. What is the term for substances that have absolutely no resistance to the flow of electricity?

    (a) hyperconductors

    (b) semiconductors

    (c) superconductors

    (d) transistors

24. A circuit runs from a battery to a lightbulb, then to another lightbulb, and finally back to the battery. What kind of circuit is this?

    (a) a double circuit

    (b) a parallel circuit

    (c) a series circuit

    (d) a unified circuit

25. How does a CD store a sound wave?

    (a) It converts the sound waves into geometric shapes.

    (b) It converts the sound wave mathematically into a series of 1's and 0's.

    (c) It creates a magnetic pattern that corresponds to the sound wave.

    (d) Its grooves contain waves that correspond to the sound wave.

26. Who formulated the Special Theory of Relativity?

    (a) Bohr

    (b) Einstein

    (c) Fermi

    (d) Planck

27. What is the term for the smallest possible amount of any substance?

    (a) an atom

    (b) a compound

    (c) a molecule

    (d) a quantum

28. If the closest star to Earth were to become a supernova at this moment, how soon would this be observable on Earth?

    (a) immediately

    (b) in 10 weeks

    (c) in 2.5 years

    (d) in 4.5 years

■ Part 1: Forces
MASTERY TEST A

Directions: Circle the correct answer to each of the following questions.

1. What term is used to describe the motion of objects?

    (a) dynamics

    (b) kinematics

    (c) mechanics

    (d) time-motion studies

2. What is translational motion?

    (a) motion in which the rate of acceleration changes

    (b) motion of an object from one place to another

    (c) motion that changes direction

    (d) spinning motion around a point

3. Which of the following is NOT a vector?

    (a) displacement

    (b) force

    (c) speed

    (d) velocity

4. What is acceleration?

    (a) how quickly an object changes direction

    (b) how quickly an object changes displacement

    (c) how quickly an object changes speed

    (d) how quickly an object changes velocity

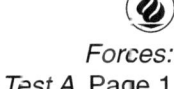

5. A runner went from 0 mph to 20 mph in 4 seconds. What was her acceleration?

   (a) 4 miles/hr/sec

   (b) 5 miles/hr/sec

   (c) 10 miles/hr/sec

   (d) 16 miles/hr/sec

6. What is the angular acceleration of the wheels of a car that is traveling at a constant velocity of 60 mph?

   (a) zero

   (b) 15 miles per hour per second

   (c) 30 miles per hour per second

   (d) 60 miles per hour per second

7. When a magician pulls a tablecloth out from under a set of plates and glasses, what keeps the plates and glasses on the table?

   (a) friction

   (b) gravity

   (c) inertia

   (d) velocity

8. I am pulling on a rope with a force of 50 pounds, and Jane is helping me with a force of 60 pounds. Maria is pulling against us with a force of 50 pounds, and Carl is helping her with a force of 70 pounds. What is the net force in this situation?

   (a) zero

   (b) 10 pounds

   (c) 110 pounds

   (d) 120 pounds

9. Misha's car has a mass of 2,000 kg and can accelerate at 5 m/sec/sec. If he puts the same engine in a car that has a mass of 1,000 kg, how fast will that car be able to accelerate? (Assume that such factors as friction and wind resistance are the same for the two cars.)

(a) 2.5 m/sec/sec

(b) 5 m/sec/sec

(c) 7.5 m/sec/sec

(d) 10 m/sec/sec

10. How much force is required to accelerate a wagon at a rate of 4 m/sec/sec whose total mass (wagon plus load) is 40 kg?

(a) 80 Newtons

(b) 120 Newtons

(c) 160 Newtons

(d) 240 Newtons

11. I want to double the torque on a rotating object while exerting the same force. Where should I now exert the force?

(a) at one-quarter the distance from the axis of rotation

(b) at one-half the distance from the axis of rotation

(c) at one and a half times the distance from the axis of rotation

(d) at twice the distance from the axis of rotation

12. Which of the following is NOT an example of Newton's Third Law?

(a) The rotors of a helicopter push the air downwards, while the air pushes the helicopter up.

(b) You crash into a wall and fall backwards.

(c) You fire a rifle, which kicks your shoulder.

(d) You press the brakes of your car to stop.

13. If you drop a rock from a high place, what will be its velocity after 10 seconds?

(a) 1 meter/sec

(b) 10 meters/sec

(c) 100 meters/sec

(d) 1,000 meters/sec

14. On Earth's surface, you are about 6,000 kilometers from Earth's center. If you were in a spaceship 12,000 kilometers above Earth's surface, what would your weight be?

    (a) one-ninth of what it was on the surface

    (b) one-quarter of what it was on the surface

    (c) four times what it was on the surface

    (d) nine times what it was on the surface

15. What would happen if the moon's horizontal motion speeded up too much?

    (a) It would be able to acquire an atmosphere.

    (b) It would fall into Earth.

    (c) It would fly away from Earth.

    (d) There would be no effect.

16. Why are astronauts weightless in space, even though the gravity of Earth and other bodies continues to pull on them?

    (a) because Earth's pull is exactly balanced by the combined pull of the sun and moon

    (b) because Earth's pull is exactly balanced by the moon's pull

    (c) because there is no air resistance

    (d) because they are in freefall

17. What is escape velocity?

    (a) the velocity at which air resistance becomes zero

    (b) the velocity of an object attracted by a black hole

    (c) the velocity that will allow an object to completely escape the gravitational pull of another object

    (d) the velocity that will put one object in orbit around another object

18. How would you describe the gravitational attraction of Earth, the moon, and the sun?

    (a) All three bodies attract one another.

    (b) Earth is attracted by the sun, but not by the moon.

    (c) The moon is attracted by Earth, but not by the sun.

    (d) The sun attracts Earth, but not the moon.

Rotational Motion

Galileo showed that gravity causes all objects to fall with constant acceleration. In this experiment, we will observe what happens when objects with different diameters roll down a slope.

Find three round objects of different sizes—such as a marble, a golf ball, and a bowling ball.

Judging from their sizes, which object do you think will be the first to reach the bottom of a slope?

_____

Now, measure the diameter of each object, and record it in the table on the right. Find a smooth slope, about two meters long. (For example, you could lean a board up against a table.) Draw a starting line 2 inches from the top of the slope. With a partner, place each object on the starting line and let the objects go at the same time. Repeat this 3 times.

Object	Diameter

What did you observe?

_____

How did the actual results compare to your predictions?

_____

_____

Can you make any conclusions about how the diameters of the objects affected your results?

_____

_____

_____

_____

Name _____

Date _____

Reaction Time

When you must react with your body to something that is happening around you, it takes only a fraction of a second for your brain to register that an event has occurred and to send the message to your muscles. This is known as your *reaction time*. In this activity, you will use Galileo's law of falling bodies to measure your reaction time.

The equation below shows the relationship between distance, time, and acceleration. Using the equation and the times provided in the chart at the right, calculate the distance the ruler falls for each given length of time. Record your answers in the chart.

$$d = \frac{1}{2}a \times t^2$$

(where $a = 1000$ cm/sec/sec)

Time (sec)	Distance (cm)
0	0
0.05	
0.1	
0.15	
0.2	
0.25	
0.3	

Now, work with a partner. Get a ruler that is marked in centimeters. Ask your partner to hold the ruler vertically. Place your fingers at the bottom of the ruler, but do not touch it. When your friend lets go of the ruler (without warning you), try to catch the ruler as quickly as possible. By observing the point on the ruler where your fingers caught the ruler, determine how far the ruler fell, and record that distance below.

Distance: _____

Find the distance in the chart above that is closest to the distance your ruler fell. What is your reaction time?

_____

Forces

Name _____

Date _____

Discovering the Planets

You have learned about the law of universal gravitation, which explains how the planets orbit around the sun. In order to learn a little more about our solar system, research one of the planets listed below. Using an encyclopedia, almanac, or Internet sources, try to find answers to the given questions. Then, write a paragraph about the characteristics of your planet.

Mercury	Venus	Earth	Mars	Jupiter
Saturn	Uranus	Neptune	Pluto	

1. How far is the planet from the sun?
2. How big is the planet?
3. What is the temperature of the planet?
4. If the planet has an atmosphere, what is it made of?
5. When you look at all the planets, how is the planet you have chosen unique?
6. Could people live on the planet?

_____

_____

_____

_____

_____

_____

_____

_____

_____

_____

Notes on Reproducible Extension Activities

Activity	Skills Applied	Product(s)
Rotational Motion	measuring applying concepts explaining observations	demonstration of rotational motion
Reaction Time	observing making simple calculations using indirect measurement	measurement of reaction time
Discovering the Planets	gathering information preparing written information	written description of a planet

Additional Extension Activity Suggestions

- Isaac Newton is often considered to be the father of modern science and technology. Discuss how his findings have affected the world we live in today. Ask your students if they think science has had a positive effect or a negative effect on the world and why.

- Ask students to think of as many everyday examples of Newton's three laws as they can. Discuss how almost any motion can be attributed to one of these laws.

- Invite learners to find out more about Galileo (or other early scientists) by looking at a Web site such as the following: http://muse.tau.ac.il/~museum/galileo/galileo.html.

Learning Modalities

Try to emphasize the similarity between translational motion and rotational motion. Although the concepts have different names, the underlying ideas are the same.

Name _____

Date _____

■ Part 1: Forces
MASTERY TEST B

Directions: Circle the correct answer to each of the following questions.

1. What is the term for explaining the causes of the motion of objects?

   (a) dynamics

   (b) kinematics

   (c) mechanics

   (d) time-motion studies

2. A train traveled for 5 hours at 70 miles per hour. What distance did it travel?

   (a) 14 miles

   (b) 75 miles

   (c) 350 miles

   (d) 570 miles

3. Malika wants to reach a town that is 150 miles away in exactly 3 hours. What must her average speed be?

   (a) 50 miles per hour

   (b) 75 miles per hour

   (c) 100 miles per hour

   (d) 150 miles per hour

4. A car is accelerating at the rate of 9 miles/hr/sec. At the end of 7 seconds what will be its velocity?

   (a) 16 mph

   (b) 32 mph

   (c) 45 mph

   (d) 63 mph

5. As a spinning top slows down, how would you describe its angular velocity and its angular acceleration?

   (a) Angular acceleration is decreasing, but angular velocity is increasing.

   (b) Angular acceleration is increasing, but angular velocity is decreasing.

   (c) Both are slowing down.

   (d) Both are speeding up.

6. Which of the following is a scalar?

   (a) acceleration

   (b) displacement

   (c) speed

   (d) velocity

7. You are traveling in a train at 70 miles per hour, and you toss a ball a couple of feet into the air. What will happen?

   (a) Air resistance will make the ball fly to the rear of the compartment.

   (b) Air resistance will make the ball return to your hand just as though you were not moving.

   (c) Inertia will make the ball fly to the rear of the compartment.

   (d) Inertia will make the ball return to your hand just as though you were not moving.

8. José's car has a mass of 1,000 kg and can accelerate at 6 m/sec/sec. He now puts the same engine in a car that has a mass of 2,000 kg. This car can now accelerate at 4 m/sec/sec. What might explain this second result?

   (a) The second car has less air resistance.

   (b) The second car has more air resistance.

   (c) The second car has more friction.

   (d) The different masses are enough to explain the different accelerations.

9. Who formulated the principle of inertia?

   (a) Democritus

   (b) Einstein

   (c) Galileo

   (d) Newton

10. In the metric system, what units measure force?

    (a) horsepower

    (b) kilograms

    (c) liters

    (d) Newtons

11. Wen's car weighs 1,500 kg, and she wants it to accelerate at 6 m/sec/sec. How much force must the engine provide?

    (a) 1,500 Newtons

    (b) 6,000 Newtons

    (c) 9,000 Newtons

    (d) 15,000 Newtons

12. When applying force to a rotating object, where must the force be applied for maximum effect?

    (a) at the axis of rotation

    (b) directly toward the axis of rotation

    (c) perpendicularly away from the axis of rotation

    (d) perpendicularly toward the axis of rotation

13. On Earth, at sea level, what is the acceleration of gravity?

    (a) 1 meter/sec/sec

    (b) 5 meters/sec/sec

    (c) 10 meters/sec/sec

    (d) 100 meters/sec/sec

14. On Earth, what prevents a feather from falling to the ground as quickly as a stone?

   (a) air resistance

   (b) gravity

   (c) its mass

   (d) its weight

15. What is a geosynchronous satellite?

   (a) a satellite that circles Earth every 90 minutes

   (b) a satellite that escapes Earth's orbit

   (c) a satellite that falls back to Earth

   (d) a satellite that remains above one point on Earth

16. What is the shape of the orbit of most planets and satellites?

   (a) a circle

   (b) an ellipse

   (c) a rounded square

   (d) a rounded triangle

17. What creates a black hole?

   (a) an area of space where there is no matter at all

   (b) the collapse of a very large star

   (c) the collapse of a very small star

   (d) the swallowing up of one star by another

18. What is the escape velocity from Earth?

   (a) 10,000 miles per hour

   (b) 15,000 miles per hour

   (c) 20,000 miles per hour

   (d) 25,000 miles per hour

■ Part 2: Energy and Heat
MASTERY TEST A

Directions: Circle the correct answer to each of the following questions.

1. What is the scientific term for the ability to do work?

   (a) action

   (b) energy

   (c) force

   (d) potential

2. Which of the following situations is NOT an example of potential energy?

   (a) an arrow flying from a bow

   (b) a compressed spring

   (c) a stone held 10 meters above the ground

   (d) a stretched elastic band

3. How much potential energy is in a 200 Newton rock that is suspended 10 meters in the air?

   (a) 10 J

   (b) 200 J

   (c) 2,000 J

   (d) 4,000 J

4. What is the principle of the conservation of energy?

   (a) that energy cannot be transformed

   (b) that kinetic energy can do more work than potential energy

   (c) that potential energy is greater than kinetic energy

   (d) that the total amount of energy in a situation does not change

5. You hold a child back on a swing and then release him. At what point is his kinetic energy the greatest?

   (a) just before you release him

   (b) at the bottom of his arc

   (c) at the top of the arc away from you

   (d) Kinetic energy is equal at all points.

6. A pendulum is rocking back and forth, and gradually comes to a stop. What happens to the energy that is lost to friction?

   (a) It is converted into heat.

   (b) It is converted into kinetic energy.

   (c) It is converted into light.

   (d) It is converted into potential energy.

7. What does the temperature of an object correspond to?

   (a) the average kinetic energy of its molecules

   (b) the average potential energy of its molecules

   (c) its density

   (d) the temperature of the surrounding air

8. On what temperature scale is absolute zero equal to −273°?

   (a) the Celsius scale

   (b) the centigrade scale

   (c) the Fahrenheit scale

   (d) the Kelvin scale

9. Although heat is generally measured in calories, what other unit can also be used to measure heat?

   (a) amps

   (b) joules

   (c) Newtons

   (d) watts

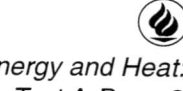

10. The specific heat of aluminum is 0.2 calories. How much heat is needed to raise the temperature of 5 grams of aluminum from 20°C to 30°C?

    (a) 1 calorie

    (b) 10 calories

    (c) 100 calories

    (d) 1,000 calories

11. After water has been warmed to 100°C, it will not actually turn to steam until additional heat has been applied. What is the term for this heat?

    (a) the latent heat of fusion

    (b) the latent heat of vaporization

    (c) the melting addendum

    (d) the phase change requirement

12. Which form of energy transfer uses a moving substance, such as water or air, to carry heat?

    (a) conduction

    (b) convection

    (c) radiation

    (d) All three forms use a moving substance to transfer heat.

13. What is the first law of thermodynamics a consequence of?

    (a) conservation of mass

    (b) Carnot efficiency

    (c) conservation of energy

    (d) entropy

14. What is the term for the study of heat and motion?

    (a) electrodynamics

    (b) hydrodynamics

    (c) mechanics

    (d) thermodynamics

15. What is Carnot efficiency?

(a) the maximum amount of heat energy that can be converted into mechanical energy

(b) the maximum percentage of heat energy that can be converted into mechanical energy

(c) the minimum amount of heat energy that can be converted into mechanical energy

(d) the minimum percentage of heat energy that can be converted into mechanical energy

16. What is the Second Law of Thermodynamics?

(a) All of the input heat can be converted into mechanical energy.

(b) Friction will always be greater than useful mechanical energy.

(c) Friction will always be less than useful mechanical energy.

(d) Not all of the input heat can be converted into mechanical energy.

17. What does the Second Law of Thermodynamics imply?

(a) Entropy always decreases.

(b) Entropy always increases.

(c) Entropy always remains constant.

(d) Entropy can neither be created nor destroyed.

18. Which of the following is an example of adiabatic compression?

(a) The temperature of a gas decreases slightly as it is compressed.

(b) The temperature of a gas decreases greatly as it is compressed.

(c) The temperature of a gas remains constant as it is compressed.

(d) The temperature of a gas rises as it is compressed.

Energy and Heat

Name _____

Date _____

Boiling Water in a Paper Cup?

Can you actually boil water in a paper cup?
Try this activity to find out.

Get a small paper cup (one without a wax
coating). Fill it three-quarters full with
water. Pick up the cup with a pair of tongs.
Hold the cup over a flame, such as one
from a Bunsen burner, an alcohol lamp, or
a gas range.

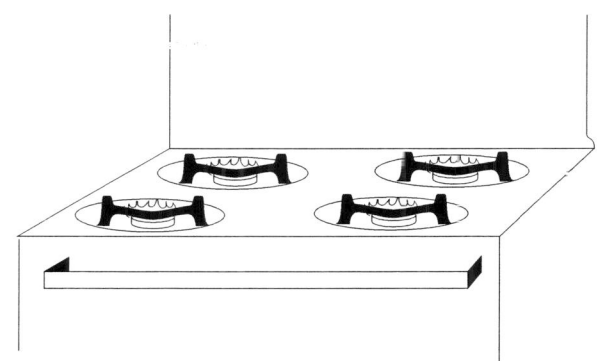

Observe that although the top of the cup
may singe, the bottom of the cup will not be
damaged. In fact, the water may even boil
in the cup.

Explain why the cup is not damaged.

_____

_____

_____

_____

_____

_____

_____

_____

_____

_____

_____

Name _____

Date _____

Heat from Food

Food contains energy. Your body releases energy when it digests food. Energy can also be released when food is burned. In this activity, you will determine the amount of energy that is contained in a food by releasing the energy and heating up water.

Fill a small beaker with 200 grams (200 ml) of water that is at room temperature. Place the beaker on a stand. Using a Celsius thermometer, measure the temperature of the water, and record it in the chart below.

Place a small nut—such as a peanut, walnut, or almond—on the end of an unfolded paper clip. Using a match, light the nut on fire. Hold the burning nut under the beaker of water so that the flame heats the water. Continue holding the nut under the beaker until the nut is completely burned. Measure the temperature of the water again, and record it in the chart below.

Then, calculate the *change* in temperature by subtracting the temperature before heating from the temperature after heating. Record your calculation in the chart.

Finally, use the equation for heat capacity to determine the amount of heat absorbed by the water. (Remember, $Q = m \times c \times T$, where $c = 1$ Cal/gm/…C for water). Record your calculation in the chart.

Temperature before heating	
Temperature after heating	
Change in temperature	
Amount of heat absorbed	

The amount of heat absorbed by the water is the amount of energy your body would gain from eating this nut!

Physics Mastery:
Energy and Heat Activity 2

Energy and Heat

Name _____

Date _____

Greenhouse Effect

Most of the energy on Earth comes from the sun by radiation. Some of this energy is absorbed by objects on Earth, and some is radiated back into space. However, certain gases can block the re-radiation of energy. This causes extra energy to accumulate on Earth, which could eventually raise the temperature of the planet. This process is known as the *greenhouse effect.*

Use newspapers, magazines, books, or the Internet to get information about the greenhouse effect. Write a short essay about what you have learned. Your essay should answer the following questions.

1. What gases are responsible for the greenhouse effect?

2. How are these gases created by man?

3. How do these gases lead to global warming?

4. What are the effects of global warming?

5. What can be done to prevent the greenhouse effect?

_____

_____

_____

_____

_____

_____

_____

_____

_____

_____

_____

_____

■ Energy and Heat

Notes on Reproducible Extension Activities

Activity	Skills Applied	Product(s)
Boiling Water in a Paper Cup	demonstrating concepts explaining observations doing hands-on activity	example of phase transition
Heat from Food	demonstrating concepts measuring making simple calculations	measurement of food energy
Greenhouse Effect	gathering information preparing written information	essay on the greenhouse effect

Additional Extension Activity Suggestions

- Discuss the different forms of energy around us and how the transformation of energy is involved in everyday processes. Emphasize how energy is useful as a concept in all areas of science.

- Ask students to list as many examples of entropy as they can. They should be able to think of many examples of things which tend to go from order into disorder. Ask them how energy can be used to reduce the entropy in each of these cases.

- Students may be interested in studying their own calorie intake. Many diet books contain tables listing the calories in various foods and the number of calories used in various physical activities. Ask students to calculate the number of calories they consume each day and the number of calories they expend.

- Ask learners to do further research into a specific type of renewable energy. A good resource is the United States Department of Energy's Web site at http://www.eren.doe.gov. This Web site contains a list of different renewable energy technologies, describing their current applications, and links to different sites that are implementing the technologies.

Pitfalls to Avoid

Many experiments in heat and energy require the use of flames or heaters. Always make sure students know the safety precautions involved in using such devices.

Name _____

Date _____

■ Part 2: Energy and Heat
MASTERY TEST B

Directions: Circle the correct answer to each of the following questions.

1. In the metric system, which unit is used to measure work?

   (a) joule

   (b) kilogram

   (c) Newton

   (d) watt

2. If a 100 kg wrecking ball has a velocity of 10 meters/sec, how much energy does it possess?

   (a) 500 J

   (b) 1,000 J

   (c) 5,000 J

   (d) 10,000 J

3. What does the principle of the conservation of matter state?

   (a) that matter can neither be created nor destroyed

   (b) that the density of matter cannot be changed

   (c) that the form of matter cannot change

   (d) that the volume of matter cannot be altered

4. You drop a rock, which is now resting on the ground. Its former potential energy has been changed into several new forms. Which of the following is NOT one of those forms?

   (a) the energy of vibrations in the ground

   (b) heat energy

   (c) the new potential energy of the rock

   (d) sound energy

5. A child is swinging back and forth on a swing, which gradually comes to a stop. What happens to the energy that is lost to air resistance?

   (a) It is converted into heat.

   (b) It is converted into kinetic energy.

   (c) It is converted into light.

   (d) It is converted into potential energy.

6. A man throws a 30 Newton ball upward with 360 J of energy. How high will it go?

   (a) 10 meters

   (b) 12 meters

   (c) 30 meters

   (d) 36 meters

7. On what temperature scale is the boiling point of water 212°?

   (a) the Celsius scale

   (b) the centigrade scale

   (c) the Fahrenheit scale

   (d) the Kelvin scale

8. How is a calorie (with a small "c") defined?

   (a) the amount of heat required to raise the temperature of 1 gram of carbon by 1°C

   (b) the amount of heat required to raise the temperature of 1 gram of iron by 1°C

   (c) the amount of heat required to raise the temperature of 1 gram of silicon by 1°C

   (d) the amount of heat required to raise the temperature of 1 gram of water by 1°C

9. After ice has been warmed to 0°C, it will not actually melt until additional heat has been applied. What is the term for this heat?

   (a) the latent heat of fusion

   (b) the latent heat of vaporization

   (c) the melting addendum

   (d) the phase change requirement

10. You have just come out of the lake on a hot day. Why do you feel cooler than if you had stayed dry?

    (a) because as water evaporates it absorbs latent heat of fusion from your body

    (b) because as water evaporates it absorbs latent heat of vaporization from your body

    (c) because as water evaporates it releases latent heat of fusion to your body

    (d) because as water evaporates it releases latent heat of vaporization to your body

11. What is the term for the simplest form of heat transfer?

    (a) conduction

    (b) convection

    (c) radiation

    (d) resistance

12. During the day, all four of the following areas were the same temperature. Assuming that the nighttime temperature was also equal in all four areas, which area would be coolest the next morning?

    (a) a dark-colored area after a clear night

    (b) a dark-colored area after a cloudy night

    (c) a light-colored area after a clear night

    (d) a light-colored area after a cloudy night

13. What happens in adiabatic expansion?

    (a) When a gas is heated, it expands.

    (b) When a gas pushes against something to expand, its temperature decreases.

    (c) When a gas pushes against something to expand, its temperature increases.

    (d) When a metal is heated, it expands.

14. What is the First Law of Thermodynamics?

    (a) Energy cannot be created or destroyed.

    (b) Heat cannot be created or destroyed.

    (c) Motion cannot be created or destroyed.

    (d) The mechanical work done by an engine cannot be greater than the heat energy supplied.

15. Which steam engine will be the most efficient?

    (a) one that heats water to 350°C and lets it cool to 75°C

    (b) one that heats water to 350°C and lets it cool to 25°C

    (c) one that heats water to 300°C and lets it cool to 75°C

    (d) one that heats water to 300°C and lets it cool to 25°C

16. What is entropy?

    (a) the amount of disorder in a system

    (b) the amount of energy in a system

    (c) the amount of friction in a system

    (d) the amount of heat in a system

17. You are using a pump to compress air in a tire. What will happen to the temperature of the pump?

    (a) It will decrease drastically.

    (b) It will decrease slightly.

    (c) It will remain constant.

    (d) It will increase.

18. Who discovered that not all of the heat energy put into an engine can be converted into mechanical energy?

    (a) Carnot

    (b) Einstein

    (c) Galileo

    (d) Newton

Name _____

Date _____

■ Part 3: Sound and Light
MASTERY TEST A

Directions: Circle the correct answer to each of the following questions.

1. What is the term for the top of a wave?

   (a) the crest

   (b) the curl

   (c) the equilibrium point

   (d) the trough

2. When calculating wavelength, where must you measure?

   (a) from crest to crest

   (b) from equilibrium point to equilibrium point

   (c) from trough to trough

   (d) between any two identical points of the wave

3. A wave hits the beach every ten seconds. What is its frequency?

   (a) 0.10 Hz

   (b) 1.0 Hz

   (c) 10.0 Hz

   (d) 100.0 Hz

4. What happens in a transverse wave?

   (a) The medium blocks the wave from traveling.

   (b) The medium vibrates in a different direction than the waves travel.

   (c) The medium vibrates in the same direction that the waves travel.

   (d) The wave travels without a medium.

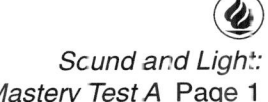

5. What happens when the crests of two different waves overlap?

    (a) constructive interference

    (b) destructive interference

    (c) maximum interference

    (d) minimum interference

6. What does the pitch of a sound wave depend upon?

    (a) frequency

    (b) loudness

    (c) speed

    (d) wavelength

7. What is the term for a bundle of light energy?

    (a) electron

    (b) neutron

    (c) photon

    (d) proton

8. Which of the following kinds of light is not really a color, but rather a combination of all colors?

    (a) blue

    (b) green

    (c) red

    (d) white

9. Which of the following is NOT one of the subtractive primary colors?

    (a) blue

    (b) cyan

    (c) magenta

    (d) yellow

10. What is the term for the set of frequencies emitted by a certain type of atom?

    (a) bandwidth

    (b) laser

    (c) prism

    (d) spectrum

11. Which of the following is true of laser light?

    (a) All its light is of the same wavelength.

    (b) It cannot be concentrated as well as ordinary light.

    (c) It contains more frequencies than ordinary light.

    (d) It is faster than ordinary light.

12. Why does red paint appear red?

    (a) because it absorbs all colors except red

    (b) because it absorbs red

    (c) because it reflects all colors except red

    (d) because it transmits all colors except red

13. When light bounces off a surface, what is the term for the angle at which it bounces off?

    (a) the angle of incidence

    (b) the angle of reflection

    (c) the angle of refraction

    (d) the normal

14. What condition is necessary for a specular reflection?

    (a) The surface must be close to room temperature.

    (b) The surface must be very hard.

    (c) The surface must be very light in color.

    (d) The surface must be very smooth.

15. Light passes from one layer of air to a slightly denser layer (which will slightly slow it down). How will the light bend?

    (a) a great deal away from the normal

    (b) a great deal toward the normal

    (c) slightly away from the normal

    (d) slightly toward the normal

16. What is the term for the point to which a converging lens brings together parallel beams of light?

    (a) cluster

    (b) "eye"

    (c) focal point

    (d) node

17. In people with normal vision, onto what part of the eye is light focused?

    (a) iris

    (b) lens

    (c) pupil

    (d) retina

18. Why does the sky appear blue on a clear day?

    (a) because the short wavelength of blue light is closest to the size of air molecules, and so blue light is scattered the most widely

    (b) because blue is the natural color of nitrogen

    (c) because blue is the natural color of oxygen

    (d) because blue light reaches Earth in higher proportions than other colors of light

Sound and Light

Name _____

Date _____

Wave Demonstration

It is easy to observe several properties of waves by creating waves along a length of rope.

Work with a partner. Get a piece of rope about 20 feet long. Ask your partner to hold the opposite end of the rope.

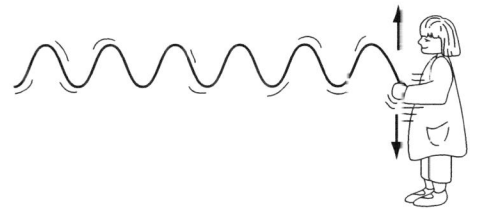

First, make low-frequency waves. Shake the rope up and down, taking about 1 second for each shake. Sketch your waves below.

Are these transverse or longitudinal waves?

_____

Next, make high frequency waves. Shake the rope up and down again, but this time shake the rope up and down three times in a second. Sketch the waves below.

Is the wavelength longer or shorter?

_____

(continued on next page)

Wave Demonstration *(continued)*

Now, make large-amplitude and small-amplitude waves. As you shake the rope, change how high you go up and how low you go down. Sketch these two types of waves below.

Large-Amplitude Wave

Small-Amplitude Wave

How would you make a large-amplitude, high-frequency wave?

_____

_____

How would you make a small-amplitude, low-frequency wave?

_____

_____

_____

Sound and Light

Name _____

Date _____

Build a Musical Instrument

In *Sound and Light,* you learned many principles about how sound is produced and how the properties of sound, such as pitch and volume, are determined. In this activity, using these principles, you will use everyday materials to design your own musical instrument.

Design and build your instrument. For example, glasses that contain different amounts of water can be tapped with a pencil and played like a xylophone. Rubber-bands of various thicknesses can be wrapped around a box to create a simple guitar.

After you have designed and built your musical instrument, answer the following questions.

What vibrates in your instrument to create the sound?

_____

_____

How can you change the amplitude of the sound?

_____

_____

How can you change the pitch of the sound?

_____

_____

Is there any common musical instrument that works on a similar principle to your instrument?

_____

_____

_____

Name _____

Date _____

Optical Devices

You have learned about the many principles that govern how light behaves. These principles have been used to design many useful devices—such as the telescope, the microscope, the camera, and the movie projector.

Go to your library and research one of these devices. Use books, magazines, and other resources to find information. Then, write a short essay that answers the following questions.

1. When and how was this device invented?

2. How does the device function? (Include a diagram if necessary.)

3. What physical principles that you learned about are involved in the function of this device?

_____

_____

_____

_____

_____

_____

_____

_____

_____

_____

_____

_____

_____

_____

_____

■ Sound and Light

Notes on Reproducible Extension Activities

Activity	Skills Applied	Product(s)
Wave Demonstration	applying concepts observing drawing conclusions	demonstration of transverse waves
Build a Musical Instrument	designing constructing applying concepts	hand-made musical instrument
Optical Devices	gathering information preparing written information applying concepts	written report on optical device

Additional Extension Activity Suggestions

- Make an impressive demonstration of the Doppler effect by tying a piece of string to a small electric buzzer (available at any electronics store). As you swing the buzzer around in a circle, the sound of the buzzer will seem higher in frequency when it is approaching the listener and lower in frequency when it is moving away.

- Many students are impressed the first time they see a laser. Inexpensive diode lasers are now available for use as pointers in presentations. A pointer can be used to demonstrate many of the concepts the students have learned, including reflection, refraction, and polarization.

- Invite learners to find out more about waves by looking at the following Web site: http//www.li.net/~stmarya/stm/home.htm. This site, created by St. Mary's High School in New York, explains different types of waves, including sound waves and the different forms of electromagnetic waves.

During the last 30 years, the use of lasers has expanded to many different fields. Today lasers are used to scan bar codes in supermarkets, to print books, to diagnose diseases, and to perform surgery. Ask your students to list as many uses of lasers as they can. Then ask them to think of possible new uses of lasers.

■ Part 3: Sound and Light
MASTERY TEST B

Directions: Circle the correct answer to each of the following questions.

1. What is the term for the substance through which a wave travels?

   (a) the aether

   (b) the bearer

   (c) the medium

   (d) the vibratory receptor

2. Which measurement describes how big a wave is?

   (a) amplitude

   (b) frequency

   (c) pulse

   (d) wavelength

3. If the frequency of a wave is 100 Hz, what is its period?

   (a) 0.01 seconds

   (b) 0.1 seconds

   (c) 1.0 second

   (d) 10.0 seconds

4. What happens in a longitudinal wave?

   (a) The medium blocks the wave from traveling.

   (b) The medium vibrates in a different direction than the waves travel.

   (c) The medium vibrates in the same direction that the waves travel.

   (d) The wave travels without a medium.

5. At which air temperature does sound travel the fastest?

    (a) 0°F

    (b) 10°F

    (c) 75°F

    (d) 100°F

6. The side of a mountain is 1500 feet away. If you shout loudly enough, how long will it take before you hear your echo?

    (a) 1 second

    (b) 1.5 seconds

    (c) 3 seconds

    (d) 15 seconds

7. What is the essential nature of light?

    (a) both a particle and a wave

    (b) neither a particle nor a wave

    (c) a particle

    (d) a wave

8. Which of the following is NOT one of the primary additive colors?

    (a) blue

    (b) green

    (c) red

    (d) yellow

9. If you mix red paint and green paint in equal proportions, and both paints are very pure, what color will you get?

    (a) black

    (b) gray

    (c) white

    (d) yellow

10. Which of the following emits coherent light?

  (a) a campfire

  (b) a laser

  (c) a lightbulb

  (d) a piece of red-hot steel

11. What sort of electromagnetic radiation is carried on optical fibers?

  (a) laser light

  (b) ordinary visible light

  (c) radio waves

  (d) X-rays

12. Which of the following is NOT one of the kinds of cones in the eye?

  (a) blue cones

  (b) green cones

  (c) red cones

  (d) yellow cones

13. When light reflects off a surface, how is the angle of incidence related to the angle of reflection?

  (a) The angle of incidence is always greater than the angle of reflection.

  (b) The angle of incidence is always equal to the angle of reflection.

  (c) The angle of incidence is always less than the angle of reflection.

  (d) This relationship depends on the smoothness of the surface.

14. What is the term for the bending of light as it passes between different transparent surfaces?

  (a) absorption

  (b) reflection

  (c) refraction

  (d) transmission

15. When light passes into a faster medium, how will it be bent?

    (a) away from the normal

    (b) in the same direction as the normal

    (c) toward the normal

    (d) It will not be bent.

16. Light passes from one layer of air to a slightly less dense layer (in which it will be able to travel slightly faster). How will the light bend?

    (a) a great deal away from the normal

    (b) a great deal toward the normal

    (c) slightly away from the normal

    (d) slightly toward the normal

17. What is the term for the different refraction of different colors?

    (a) convergence

    (b) dispersion

    (c) gathering

    (d) scattering

18. All the waves in a beam of light are precisely horizontal. What is the term for this sort of light?

    (a) diffracted

    (b) ionized

    (c) iridescent

    (d) polarized

■ Part 4: Electricity, Magnetism, and Beyond
MASTERY TEST A

Directions: Circle the correct answer to each of the following questions.

1. Which unit measures current?

    (a) the ampere

    (b) the coulomb

    (c) the volt

    (d) the watt

2. Which sort of circuit allows electricity to flow?

    (a) a blocked circuit

    (b) a closed circuit

    (c) an open circuit

    (d) a variable circuit

3. What objects do not allow electricity to flow through them easily?

    (a) conductors

    (b) insulators

    (c) resisters

    (d) transmitters

4. If a lightbulb has a resistance of 300 and you connect it to a 3 V battery, how much current will flow?

    (a) .01 A

    (b) .03 A

    (c) 1 A

    (d) 3 A

5. What problem can arise when a circuit is connected without a load?

    (a) no current

    (b) no voltage

    (c) a short circuit

    (d) too much resistance

6. You have created an electromagnet by running a current through a looped wire. Which of the following will make the magnetic field weaker?

    (a) decreasing the resistance of the circuit

    (b) increasing the current

    (c) decreasing the current

    (d) increasing the number of loops

7. A circuit runs from a battery to a lightbulb and then back to the battery. Another circuit runs from the same battery to a second lightbulb and then back to the battery again. What happens if one of the lightbulbs burns out?

    (a) The current will no longer flow.

    (b) There will be a short circuit.

    (c) The other bulb will also burn out.

    (d) The other bulb will continue to glow.

8. A wire carrying electrical current will feel a force near a magnet. The more current in the wire, the greater the force. What is the term for this phenomenon?

    (a) the dynamo effect

    (b) the generator effect

    (c) the magnetic effect

    (d) the motor effect

9. In a generator, what is the term for the large rotating shaft to which the coil of wire is attached?

   (a) the brushes

   (b) the coil

   (c) dynamo

   (d) the turbine

10. Which of the following creates direct current?

   (a) a battery

   (b) a generator

   (c) both (a) and (b)

   (d) neither (a) nor (b)

11. What does a loudspeaker do?

   (a) amplifies a sound wave into a louder sound wave

   (b) converts sound waves into a voltage

   (c) converts a voltage into a sound wave

   (d) records a sound wave

12. Even when properly handled, which of the following wears out fastest?

   (a) cassette tapes

   (b) CDs

   (c) vinyl records

   (d) All three wear out at about the same rate.

13. You are traveling in a spaceship at ¾ the speed of light, and you shine a beam of light directly behind you. To a stationary observer, how fast will the beam of light appear to be traveling?

   (a) ¼ the speed of light

   (b) ¾ the speed of light

   (c) the speed of light

   (d) 1¾ the speed of light

14. Which of the following statements about relativity is NOT true?

    (a) The rules of Special Relativity are easily observed in everyday objects.

    (b) The speed of light is the same, no matter how fast you are moving relative to the light.

    (c) It is impossible to say that an object is absolutely stationary.

    (d) All motion is relative.

15. What determines the paths of electrons around a nucleus?

    (a) the fundamental quantum of energy

    (b) interaction with other atoms

    (c) magnetism

    (d) the number of neutrons in the nucleus

16. What is the term for the force that holds protons and neutrons together in the nucleus of an atom?

    (a) electromagnetism

    (b) gravity

    (c) the strong force

    (d) the weak force

17. Which of the following is a form of electromagnetic wave?

    (a) an alpha ray

    (b) a beta ray

    (c) a gamma ray

    (d) all of the above

18. Which of the following consists of electrons that are emitted when a neutron breaks apart?

    (a) an alpha ray

    (b) a beta ray

    (c) a gamma ray

    (d) all of the above

Name _____

Date _____

Electromagnet

It is easy to build an electromagnet using materials purchased in a hardware store.

Your instructor will provide some thin, insulated wire; a nail; and two 6-volt batteries (with coil terminals on top).

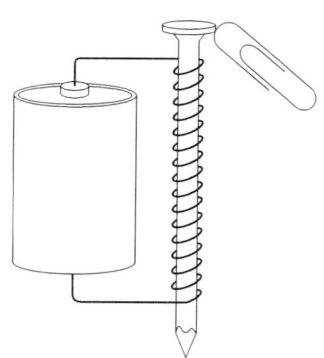

First, cut about 3 feet of wire, and remove about an inch of insulation on each end. Wrap the wire around the nail many times, leaving about six inches free on each end.

Connect the free ends to the battery. The nail should now behave like a magnet. Test its strength by measuring the number of paper clips that it can pick up. How many paper clips were picked up?

_____

Next, make another electromagnet, this time using 6 feet of wire. How many paper clips can you pick up now?

_____

Finally, try connecting your electromagnet to two batteries. Connect one end of the wire to the (+) terminal of one battery. Connect a short wire from the (−) terminal of that battery to the (+) of the second battery. Finally, connect the second end of the electromagnet to the remaining (−) terminal. How can many paper clips can you pick up now?

_____

What can you conclude about the effect of the number of loops of wire on the strength of the electromagnet?

_____

What can you conclude about the effect of the amount of voltage applied on the strength of the electromagnet?

_____

Electricity, Magnetism, and Beyond	Name _____
	Date _____

Making a Compass

You can make your own compass using a needle, a cork, a bowl of water, and a magnet.

Stroke the needle with the magnet. Be sure to rub the needle in only one direction.

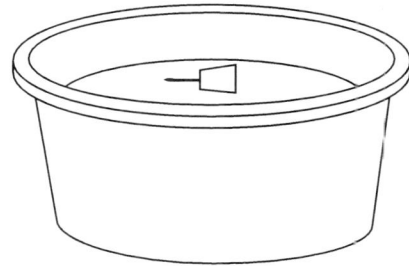

Push the needle through the cork. Fill a bowl with water, and place the cork with the needle in the water. The needle should spin until one end faces north.

What happened to the needle when you rubbed the magnet over it?

_____

_____

_____

_____

Why does the needle point north?

_____

_____

_____

_____

Semiconductors and Superconductors

Semiconductors and superconductors are two materials that are very important in modern technology.

Go to your library or search the Internet for information on either semiconductors or superconductors. Write a short essay about the material you choose.

Your essay should answer the following questions.

1. When and how was this material discovered?

2. What are the types of substances that this material is made from?

3. What are some current applications of this material?

4. What are some current areas of research about, or possible future improvements for, this material?

_____

_____

_____

_____

_____

_____

_____

_____

_____

_____

_____

_____

_____

Notes on Reproducible Extension Activities

Activity	Skills Applied	Product(s)
Electromagnet	constructing observing demonstrating concepts	electromagnet
Making a Compass	constructing observing explaining observations	compass
Semiconductors and Superconductors	gathering information preparing written information analyzing technology	essay

Additional Extension Activity Suggestions

- Invite learners to explore the Web site of the Boston Museum of Science (http://www.mos.org). The theater of electricity teaches students about static and current electricity through text, dramatic movies, and photos of the world's largest insulated DeGraph generator.

- If you can obtain a very strong horseshoe magnet, it is possible to demonstrate the motor and generator effects directly. When a length of wire is connected to a battery and placed between the two poles, students will be able to feel the magnetic force acting on the wire. Similarly, if you connect a length of wire to a sensitive ammeter, and move the wire between the two poles, the ammeter will show the induced current.

- Learners might be interested in visiting the online Holography Science Museum (http://www.holoworld.com). Frank DeFreitas, who owns a hologram gallery in Allentown, PA, has set up an award-winning Internet Webseum of Holography. There students will learn how lasers work and how holograms are made. They can also view laser light shows, demonstrations of optic principles, and laser experiments.

Fascinating Facts

An electric eel has current-producing organs that can release a charge of 350–550 volts (on average). The most powerful electric eels, found in the rivers of Brazil, Colombia, Venezuela, and Peru, produce a shock of 400–650 volts.

■ Part 4: Electricity, Magnetism, and Beyond
MASTERY TEST B

Directions: Circle the correct answer to each of the following questions.

1. Which unit measures electric charge?

   (a) the ampere

   (b) the coulomb

   (c) the volt

   (d) the watt

2. Which objects allow electricity to flow through them easily?

   (a) conductors

   (b) insulators

   (c) resisters

   (d) transmitters

3. Which unit measures resistance?

   (a) the ampere

   (b) the ohm

   (c) the volt

   (d) the watt

4. Which part of the circuit uses electricity?

   (a) the burden

   (b) the charge

   (c) the load

   (d) the resistor

5. What is the role of fuses or circuit breakers?

   (a) to cut off the voltage supply when too much current is flowing

   (b) to detect when current is too weak

   (c) to supply extra current

   (d) to supply extra voltage

6. If you hammer an iron magnet, what will happen to its magnetism?

   (a) It will act in a different direction.

   (b) It will decrease.

   (c) It will increase.

   (d) It will not be affected.

7. A circuit runs from a battery to a lightbulb and then back to the battery. Another circuit runs from the same battery to a second lightbulb and then back to the battery again. What kind of circuit is this?

   (a) a double circuit

   (b) a parallel circuit

   (c) a series circuit

   (d) a unified circuit

8. What term describes the process in which a magnet is used to create a current in a moving wire?

   (a) the dynamo effect

   (b) the generator effect

   (c) the magnetic effect

   (d) the motor effect

9. Which of the following creates alternating current?

   (a) a battery

   (b) a generator

   (c) both

   (d) neither

10. What does a microphone do?

    (a) amplifies a sound wave into a louder sound wave

    (b) converts sound waves into a voltage

    (c) converts a voltage into a sound wave

    (d) records a sound wave

11. How does a vinyl record store a sound wave?

    (a) It converts the sound waves into geometric shapes.

    (b) It converts the sound waves mathematically into a series of 1's and 0's.

    (c) It creates a magnetic pattern that corresponds to the sound wave.

    (d) Its grooves contain waves that correspond to the sound wave.

12. What does an ammeter measure?

    (a) the amount of current in a wire

    (b) the amount of resistance in a wire

    (c) the amount of voltage in a wire

    (d) the size of the load in a circuit

13. If you traveled in a spaceship close to the speed of light for 10 years, how much time would have passed on Earth when you returned?

    (a) 5 years

    (b) 10 years

    (c) 15 years

    (d) 20 years

14. Why do we never notice quantization?

    (a) because the fundamental quanta are so large

    (b) because the fundamental quanta are so small

    (c) because we are moving quickly relative to such objects

    (d) because we are moving slowly relative to such objects

15. How do very small objects behave?

    (a) like particles

    (b) like waves

    (c) like both waves and particles

    (d) like neither waves nor particles

16. What is the term for the force that holds quarks together to form protons and neutrons?

    (a) electromagnetism

    (b) gravity

    (c) the strong force

    (d) the weak force

17. Which of the following has the greatest penetrating power?

    (a) an alpha ray

    (b) a beta ray

    (c) a gamma ray

    (d) all of the above

18. Which of the following is one of the leading areas of research in astronomy?

    (a) the search for UFOs

    (b) the search for visible matter

    (c) the search for light matter

    (d) the search for dark matter

Name _____

Date _____

■ Physics Mastery
COURSE MASTERY TEST B

Directions: Circle the correct answer to each of the following questions.

1. What do you need to know to determine the displacement of an object?

   (a) the direction an object has moved

   (b) the distance an object has moved

   (c) the direction and distance an object has moved

   (d) the distance and speed an object has moved

2. What is the term for the imaginary line around which an object spins?

   (a) the angular displacement

   (b) the angular velocity

   (c) the axis of rotation

   (d) the axis of displacement

3. When a car makes a quick turn, what acts on the passengers to push them toward the outside?

   (a) friction

   (b) gravity

   (c) inertia

   (d) velocity

4. Four people, all wearing roller skates, throw the same size basketball. Each of them is pushed backward in reaction. Who will move the least?

   (a) Aparna, who weighs 50 kilograms

   (b) Julia, who weighs 55 kilograms

   (c) Hiroshi, who weighs 70 kilograms

   (d) Jamal, who weighs 80 kilograms

5. Who formulated the Law of Falling Objects?

   (a) Copernicus

   (b) Einstein

   (c) Galileo

   (d) Newton

6. If you drop a brick from a high place, how far will it fall in 10 seconds?

   (a) 50 meters

   (b) 100 meters

   (c) 500 meters

   (d) 1000 meters

7. What is the event horizon of a black hole?

   (a) the diameter of a black hole

   (b) the distance from a black hole that marks the point at which nothing can escape

   (c) the event that created the black hole

   (d) the time when the black hole was created

8. Al lifts a 3 kg mass 1 meter; Sumi lifts 2 kg 1.5 meters; and Sarah lifts 1 kg 3 meters. Who has done the most work?

   (a) Al

   (b) Sumi

   (c) Sarah

   (d) They have all done the same amount of work.

9. You hold a child back on a swing and then release him. At what point is his potential energy the greatest?

   (a) just before you release him

   (b) at the bottom of his arc

   (c) at the top of the arc away from you

   (d) Potential energy is equal at all points.

10. What causes thermal expansion?

    (a) As molecules move faster, they jostle one another farther apart.

    (b) As molecules move faster, they grow in length.

    (c) As molecules move slower, they jostle one another farther apart.

    (d) As molecules move slower, they grow in length.

11. What is specific heat?

    (a) the ability of an object to exchange heat with another object

    (b) the amount of heat required to raise the temperature of 1 gram of a substance by 1°C

    (c) the temperature at which a substance becomes a gas

    (d) the temperature at which a substance melts

12. Which of the following forms of energy transfer uses electromagnetic waves?

    (a) conduction

    (b) convection

    (c) radiation

    (d) All three forms use electromagnetic radiation.

13. Which of the following is NOT an example of conduction?

    (a) It is easy to burn your hand on a metal spoon in a pan of boiling water.

    (b) Your hand feels cold when you hold a snowball.

    (c) You fry an egg in a metal pan.

    (d) You dry your clothes by placing them in a sunny location.

14. Which of the following is TRUE of entropy?

    (a) Entropy always increases.

    (b) Entropy never changes.

    (c) Entropy always decreases.

    (d) Entropy has a positive effect on the environment.

15. What is the term for the bottom of a wave?

   (a) the crest

   (b) the curl

   (c) the equilibrium point

   (d) the trough

16. What do decibels measure?

   (a) the frequency of sound

   (b) the loudness of sound

   (c) the speed of sound

   (d) the wavelength of sound

17. Through which of the following does light travel fastest?

   (a) empty space

   (b) glass

   (c) optical fibers

   (d) water

18. According to modern physics, all objects behave as waves sometimes and as particles sometimes. Under what condition does an object act more like a wave?

   (a) the colder it is

   (b) the hotter it is

   (c) the larger it is

   (d) the smaller it is

19. What is the term for what happens when a ray of light strikes an object and bounces off?

   (a) absorption

   (b) reflection

   (c) refraction

   (d) transmission

20. What is the focal length of a lens?

    (a) the distance between the eye and the lens

    (b) the distance between the eye and the object to be seen

    (c) the distance between the focal point and the lens

    (d) the distance between the focal point and the object to be seen

21. Why does the sun appear red at sunset?

    (a) because the sunlight has to travel through less of the atmosphere, and the long wavelengths of red light scatter least

    (b) because the sunlight has to travel through less of the atmosphere, and the long wavelengths of red light scatter most

    (c) because the sunlight has to travel through more of the atmosphere, and the long wavelengths of red light scatter least

    (d) because the sunlight has to travel through more of the atmosphere, and the long wavelengths of red light scatter most

22. What is the term for the flow of electrons?

    (a) charge

    (b) current

    (c) electric potential

    (d) voltage

23. What will happen if you cut a bar magnet in two, halfway between its north pole and its south pole?

    (a) You will destroy the magnetism in both halves.

    (b) You will have two magnets, but there will no longer be north and south poles.

    (c) You will have two magnets, each with both a north pole and a south pole.

    (d) You will have two magnets, one with only a south pole and one with only a north pole.

24. A circuit runs from a battery to a lightbulb, then to another lightbulb, and finally back to the battery. What happens if one of the lightbulbs burns out?

    (a) The current will no longer flow.

    (b) Nothing will happen.

    (c) The other bulb will glow half as brightly.

    (d) The other bulb will glow twice as brightly.

25. In a CD player, what sort of light is reflected from the surface of the disk?

    (a) infrared light

    (b) laser light

    (c) ordinary visible light

    (d) polarized light

26. You are on a train traveling 75 mph. Standing at the head of the train, you throw a ball to a friend at the back at a speed of 40 mph. To a person on the ground, outside the train, how fast would the ball be moving?

    (a) 35 mph

    (b) 40 mph

    (c) 75 mph

    (d) 115 mph

27. Quantum mechanics describes the strange consequences of what kind of objects?

    (a) objects that are at very low temperatures

    (b) objects that are traveling at speeds approaching the speed of light

    (c) objects that are very large

    (d) objects that are very small

28. The half-life of carbon-14 is about 6,000 years. If you begin with one ounce of this material, how much of it will still be carbon-14 after 12,000 years?

    (a) ⅛ ounce

    (b) ¼ ounce

    (c) ½ ounce

    (d) ¾ ounce

Answer Key

PHYSICS MASTERY

Course Mastery Test A

1. c 2. d 3. d 4. b 5. d 6. b 7. a 8. c 9. d 10. c 11. b 12. c 13. c 14. b
15. d 16. c 17. d 18. c 19. a 20. c 21. c 22. b 23. c 24. c 25. b 26. b 27. d
28. d

Forces: Mastery Test A

1. b 2. b 3. c 4. d 5. b 6. a 7. c 8. b 9. d 10. c 11. d 12. d 13. c 14. a 15. c
16. d 17. c 18. a

Forces: Mastery Test B

1. a 2. c 3. a 4. d 5. c 6. c 7. d 8. a 9. d 10. d 11. c 12. c 13. c 14. a 15. d
16. b 17. b 18. d

Energy and Heat: Mastery Test A

1. b 2. a 3. c 4. d 5. b 6. a 7. a 8. d 9. b 10. b 11. b 12. b 13. c 14. d
15. b 16. d 17. b 18. d

Energy and Heat: Mastery Test B

1. a 2. c 3. a 4. c 5. a 6. b 7. c 8. d 9. a 10. b 11. a 12. c 13. b 14. d 15. b
16. a 17. d 18. a

Sound and Light: Mastery Test A

1. a 2. d 3. a 4. b 5. a 6. a 7. c 8. d 9. a 10. d 11. a 12. a 13. b 14. d
15. d 16. c 17. d 18. a

Sound and Light: Mastery Test B

1. c 2. a 3. a 4. c 5. d 6. c 7. a 8. d 9. a 10. b 11. a 12. d 13. b 14. c 15. a
16. c 17. b 18. d

Electricity, Magnetism, and Beyond: Mastery Test A

1. a 2. b 3. b 4. a 5. c 6. c 7. d 8. d 9. d 10. a 11. c 12. c 13. c 14. a 15. a
16. c 17. c 18. b

Electricity, Magnetism, and Beyond: Mastery Test B

1. b 2. a 3. b 4. c 5. a 6. b 7. b 8. b 9. b 10. b 11. d 12. a 13. d 14. b
15. c 16. d 17. c 18. d

Course Mastery Test B

1. c 2. c 3. c 4. d 5. c 6. c 7. b 8. d 9. a 10. a 11. b 12. c 13. d 14. a 15. d
16. b 17. a 18. d 19. b 20. c 21. c 22. b 23. c 24. a 25. b 26. a 27. d 28. b

ANSWER KEY

PART 1: FORCES

Lesson 1: Kinematics

Page 7:	**1.** b **2.** c **3.** c **4.** a **5.** c **6.** a
Page 9:	**1.** b **2.** c **3.** a **4.** a
Page 12:	**1.** b **2.** c **3.** c **4.** b
Page 14:	**Think About It:** This means the length of a day on Earth is slightly increasing. Of course, this change is very slow. The day only changes by one second over thousands of years. But, if you've ever wished you had more time in the day, just wait a few million years!
Page 15:	**1.** c **2.** a **3.** b **4.** b **5.** b

Lesson 1: Lesson Mastery Test

Page 16:	**1.** R **2.** T **3.** T **4.** R **5.** V **6.** D **7.** A **8.** S **9.** a **10.** b **11.** c **12.** b **13.** b

Lesson 2: Dynamics

Page 19:	**Think About It:** No. In outer space, there is no air, so there is no friction. Therefore, the inertia of the spaceship will allow it to continue moving in a straight line at the same speed. In fact, several unmanned spacecraft have left the solar system, and will continue to move by inertia for thousands or millions of years.
Page 20:	**1.** b **2.** c **3.** b **4.** F **5.** F **6.** T
Page 24:	**1.** a **2.** b **3.** c **4.** c **5.** c **6.** c
Page 26:	**1.** a **2.** b **3.** c **4.** b
Page 27:	**Think About It:** (a) The air pushes the helicopter up. (b) The barbell pushes down on the weight-lifter. (c) The telephone pole pushes back on the car, making it stop.
Page 28:	**Think About It:** The small child will. They both feel the same reaction force. However, the small child has less mass, so she will accelerate more.
Page 28:	**1.** c **2.** b **3.** b **4.** T **5.** T **6.** F
Page 32:	**1.** b **2.** a **3.** b **4.** T **5.** F **6.** T **7.** T

Lesson 2: Lesson Mastery Test

Page 33:	**1.** B **2.** C **3.** A **4.** B **5.** C **6.** b **7.** b **8.** b **9.** b **10.** c **11.** c

Lesson 3: Gravity

Page 38:	**1.** c **2.** b **3.** c **4.** a **5.** c **6.** b
Page 41:	**1.** b **2.** b **3.** a **4.** c **5.** F **6.** F **7.** T
Page 44:	**Think About It:** If there were no gravity, they would not be in orbit! The Space Shuttle would float away from Earth in a straight line. Earth's gravity pulls the Space Shuttle into a circular orbit.
Page 45:	**1.** F **2.** T **3.** T **4.** T **5.** T **6.** F **7.** T
Page 48:	**1.** b **2.** b **3.** a **4.** F **5.** T **6.** F **7.** F **8.** T **9.** T

Lesson 3: Lesson Mastery Test

Page 49:	**1.** b **2.** e **3.** d **4.** c **5.** a **6.** a **7.** b **8.** b **9.** c **10.** c **11.** F **12.** T **13.** T **14.** F **15.** F

ANSWER KEY

PART 2: ENERGY AND HEAT

Lesson 1: Mechanical Energy

Page 55: **1.** c **2.** a **3.** a **4.** b **5.** c

Page 57: **Think About It:** You can increase the amount of force by making the bow tighter. The more difficult it is to pull the string back, the more force you will need. Therefore, you will store more energy. You can also pull the arrow back farther. By increasing the distance over which the force is applied, you do more work. This also stores more energy.

Page 59: **1.** K **2.** G **3.** K **4.** E **5.** F **6.** T **7.** F **8.** F **9.** b **10.** a

Page 63: **1.** F **2.** T **3.** T **4.** F **5.** T **6.** T

Page 66: **1.** 250 J **2.** 250 J **3.** 25 N **4.** 150 J **5.** 150 J **6.** 0.2 m

Lesson 1: Lesson Mastery Test

Page 67: **1.** K **2.** E **3.** G **4.** K **5.** E **6.** c **7.** b **8.** b **9.** F **10.** F **11.** T **12.** F
13. 140 J **14.** 7 meters

Lesson 2: Temperature and Heat

Page 72: **1.** a **2.** a **3.** c **4.** a **5.** a **6.** b **7.** c

Page 76: **1.** b **2.** c **3.** c **4.** a **5.** b **6.** F **7.** T **8.** F

Page 80: **1.** F **2.** F **3.** T **4.** T **5.** F **6.** T **7.** F **8.** T

Page 82: **Think About It:** The air, cake, and metal all conduct heat differently. Air is a poor conductor of heat, so it transfers very little heat into your hand. The cake is a moderate conductor of heat, so it feels somewhat hot to the touch. Metal is such a good conductor that heat is transferred into your hand quickly enough to burn you before you can pull back.

Page 85: **1.** A **2.** C **3.** B **4.** C **5.** A **6.** B **7.** F **8.** T **9.** T **10.** T

Lesson 2: Lesson Mastery Test

Page 86: **1.** a **2.** a **3.** b **4.** c **5.** F **6.** F **7.** F **8.** T **9.** T **10.** B **11.** A **12.** B
13. C **14.** A

Lesson 3: Heat Engines and Thermodynamics

Page 90: **1.** C **2.** E **3.** E **4.** F **5.** T **6.** T **7.** T **8.** F

Page 93: **1.** b **2.** a **3.** a **4.** b **5.** F **6.** T

Page 96:   **1.** F   **2.** F   **3.** F   **4.** T   **5.** T

Lesson 3: Lesson Mastery Test

Page 97:   **1.** b   **2.** b   **3.** b   **4.** c   **5.** F   **6.** T   **7.** T   **8.** F

ANSWER KEY
PART 3: SOUND AND LIGHT

Lesson 1: Sound Waves

Page 103:	**1.** b **2.** a **3.** c **4.** c **5.** a **6.** b
Page 107:	**1.** F **2.** T **3.** T **4.** F **5.** T **6.** T **7.** F **8.** F **9.** F
Page 111:	**1.** T **2.** F **3.** T **4.** T **5.** F **6.** T **7.** T **8.** F **9.** F
Page 113:	**Think About It:** The high notes are the short, thin strings. They have a short wavelength, which means high natural frequency. They are also thin, so they vibrate quickly.
Page 114:	**1.** c **2.** c **3.** b **4.** b **5.** a **6.** a
Page 117:	**1.** b **2.** a **3.** c **4.** c

Lesson 1: Lesson Mastery Test

Page 118:	**1.** a **2.** b **3.** c **4.** a **5.** b **6.** b **7.** c **8.** b **9.** b **10.** a **11.** T **12.** F **13.** T **14.** T **15.** F **16.** F

Lesson 2: The Nature of Light

Page 122:	**1.** c **2.** b **3.** a **4.** c **5.** c
Page 125:	**1.** b **2.** a **3.** a **4.** a **5.** c
Page 129:	**1.** a **2.** a **3.** a **4.** c **5.** b **6.** F **7.** T **8.** F

Lesson 2: Lesson Mastery Test

Page 130:	**1.** c **2.** b **3.** c **4.** a **5.** a **6.** F **7.** T **8.** T **9.** T **10.** F **11.** T

Lesson 3: The Behavior of Light

Page 133:	**1.** c **2.** a **3.** b **4.** a **5.** b
Page 137:	**1.** b **2.** a **3.** b
Page 138:	**4.** a **5.** a **6.** b **7.** b **8.** a **9.** b

Page 140:   **Think About It:** Think of the shaking of a rope through the slot in a fence as you try to figure this out. Suppose the first pair of sunglasses is polarized vertically. It lets only vertically polarized light through. If the second set of glasses has the same polarization, there would be no change. The light could still get through the second pair. However, if the second pair were polarized horizontally, the light could not get through. You would not be able to see through them. If the second pair were held at an in-between angle, some of the light could pass through, but some would be absorbed.

Page 141:   **1.** F   **2.** F   **3.** T   **4.** T   **5.** F   **6.** T   **7.** T

Page 143:   **Think About It:** The molecules from the milk can scatter light just like the air molecules in the atmosphere. When you look from the side, you are seeing the scattered light which is blue. This is like looking up at the sky. When you look from the opposite end, you are seeing the light which makes it through without being scattered. This is the red light. This is like looking at the light of the setting sun. It must pass through a lot of air before reaching your eyes.

Page 144:   **1.** c   **2.** a   **3.** c   **4.** b   **5.** a

Lesson 3: Lesson Mastery Test

Page 145:   **1.** a   **2.** b   **3.** b   **4.** a   **5.** a   **6.** b   **7.** a   **8.** c   **9.** F   **10.** F   **11.** T   **12.** F
              13. T   **14.** T   **15.** T

PART 4: ELECTRICITY, MAGNETISM, AND BEYOND

Lesson 1: The Basis of Electricity and Magnetism

Page 149:   **Think About It :** Answers will vary. Common examples include rubbing your feet on a carpet, rubbing a balloon across a sweater, or clothing that sticks together in the dryer.

Page 150:   **1.** a   **2.** b   **3.** b   **4.** a

Page 153:   **1.** b   **2.** b   **3.** a   **4.** b   **5.** b   **6.** b   **7.** c   **8.** a   **9.** b

Page 156:   **1.** F   **2.** F   **3.** T   **4.** F   **5.** T   **6.** F   **7.** T

Page 160:   **1.** F   **2.** T   **3.** T   **4.** T   **5.** F   **6.** F   **7.** T   **8.** T

Lesson 1: Lesson Mastery Test

Page 161:   **1.** b   **2.** a   **3.** d   **4.** c   **5.** a   **6.** b   **7.** a   **8.** b   **9.** b   **10.** a   **11.** T   **12.** T
             13. F   **14.** F   **15.** F   **16.** F   **17.** T   **18.** T

Lesson 2: Applications of Electricity and Magnetism

Page 165:   **Think About It:** It is best if they are connected in parallel, so that if one lightbulb burns out, the rest can continue to light up. However, it requires much more wire to connect the bulbs in parallel, so inexpensive strings of holiday lights are often wired in series.

Page 165:   **1.** b   **2.** b   **3.** b   **4.** a   **5.** a

Page 168:   **1.** F   **2.** F   **3.** T   **4.** F   **5.** F   **6.** T

Page 171:   **1.** F   **2.** T   **3.** F   **4.** T   **5.** F   **6.** T   **7.** F

Page 175:   **1.** T   **2.** T   **3.** F   **4.** T   **5.** D   **6.** C   **7.** R   **8.** C

Lesson 2: Lesson Mastery Test

Page 176:   **1.** h   **2.** e   **3.** b   **4.** g   **5.** a   **6.** d   **7.** i   **8.** a   **9.** a   **10.** a   **11.** T   **12.** T   **13.** F
             14. T   **15.** T   **16.** T

Lesson 3: Modern Physics

Page 179:   **Think About It:** As surprising as it may seem, you both see the same speed: 186,000 miles per second.

Page 181:   **1.** F   **2.** T   **3.** T   **4.** T   **5.** T   **6.** F   **7.** F   **8.** F   **9.** F

Page 184:   **1.** T   **2.** T   **3.** F   **4.** F   **5.** T   **6.** T

Page 188:   **1.** a   **2.** c   **3.** a   **4.** c   **5.** c   **6.** c

Page 191:   **1.** F **2.** T **3.** T **4.** T **5.** F **6.** T **7.** F **8.** F

Lesson 3: Lesson Mastery Test

Page 192:   **1.** C **2.** R **3.** Q **4.** Q **5.** RA **6.** R **7.** C **8.** RA **9.** b **10.** b **11.** a **12.** b
                 13. F **14.** T **15.** F **16.** T **17.** T **18.** T **19.** T

Reproducible
Student
Pages

To the Student

Welcome to *Physics Mastery*, a course designed to help you understand the basic concepts of physics.

Part One, Forces teaches about forces and how they affect the motion of objects. You will learn about gravity, kinematics, and dynamics.

In **Part Two, Energy and Heat,** you will learn about the various processes involving the transfer of heat, the laws of thermodynamics, and heat engines.

Part Three, Sound and Light teaches the terms and principles of sound waves, the nature of light, and how light behaves when interacting with matter around us.

In **Part Four, Electricity, Magnetism, and Beyond,** you will learn about some of the most recent developments in physics, the applications of electricity and magnetism, and the beginning of the universe.

Each lesson in the book includes many things to help you learn. "Tips" will give you hints on how to make learning easier. "In Real Life" sections will show you how the skills you are learning can be practiced and used every day. "Think About It" questions will ask you to think about the physical world in new ways. Finally, every lesson ends with a lesson mastery test. These tests will help you check what you have just learned before you go on to new material.

We hope you will use *Physics Mastery* to master skills that will help you understand and appreciate the world around you. And, we hope you will enjoy yourself as you learn!

PART 1

FORCES

Table of Contents

■ Lesson 1—Kinematics

Goal: To understand the terms and equations used to describe motion

Translation and Vectors

Compare the following statements.

The car moved northeast at the rate of 45 miles per hour.

The car moved because the turning wheels applied a force to the body of the car.

The first statement describes *how* the car moved, and the second statement explains *why* the car moved. Describing the motion of objects is known as **kinematics.** Explaining the causes of motion is known as **dynamics.** Kinematics and dynamics are two branches of the overall study of motion, known as **mechanics.**

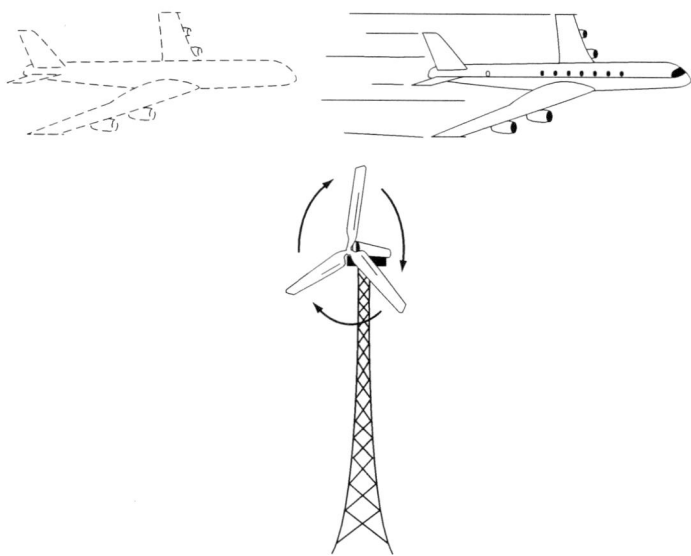

The diagram above illustrates two types of motion: (1) the motion of an airplane, and (2) the motion of a windmill.

The motion of an object, such as an airplane, from one place to another is known as **translational** motion. When an object, such as a windmill, spins around in one place it is known as **rotational** motion. Some objects, such as a rolling bowling ball, exhibit both translational and rotational motion.

This lesson will explain the kinematics of translational motion.

Suppose an object is located at position A in Figure 1 below, and you are told that the object moved one inch. You don't know whether it is now at position B, C, or D, or even somewhere else.

Now, suppose you are told that the object in Figure 2 is moved from position A toward the upper right-hand corner. Again, you don't know exactly which position it is in. The object could be at position B, C, D, or at some other point.

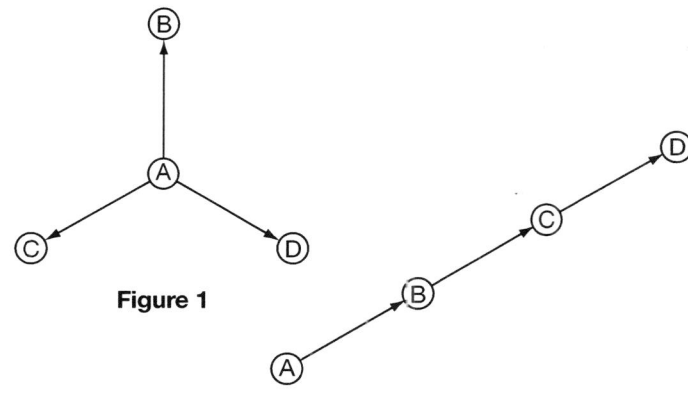

Figure 1

Figure 2

In the first situation, you are told the **distance** the object has moved. In the second case, you are told the **direction** the object has moved. In order to completely describe its motion, you need to give both the direction *and* distance it moved. Together, these are known as the **displacement** of the object. If you say the object moved one inch toward the upper right-hand corner, you have exactly described its new location.

Similarly, if you want to describe an object that is still moving, it is not enough just to say, "It is moving 20 miles per hour." This describes the object's **speed.** But, if you do not know the direction it is moving, you cannot predict where it will be some time later. You need to know the speed *and* direction of the object. Speed and direction together are known as the **velocity** of the object.

It is important to understand the difference between speed and velocity. A car can change its velocity without changing its speed. How is that possible? When a car changes direction, it changes velocity. A car going 30 mph north, and another car going 30 mph south, have the same speed, but opposite velocities.

Quantities such as velocity and displacement, which combine a number and a direction, are known as **vectors.** Vectors tell you "how much," and "in what direction." For example, displacement tells you how far and in what direction. Velocity tells you how fast and in what direction. Force is also a vector. In order to describe a push on something, you must say how hard you pushed, and in what direction. The "how much" of a vector is known as the **magnitude.** For example, the magnitude of velocity is its speed. The magnitude of a force is how strong the force is.

A vector is often indicated by an arrow. The length of the arrow indicates the magnitude, and the point of the arrow indicates the direction. In the diagram below, for example, you can tell by the length of the arrows that the car heading east on Pine Street is moving faster than the car heading north on Elm Street. The longer arrow on the Pine Street car shows that it is moving with a greater magnitude (that is, with more force, or speed).

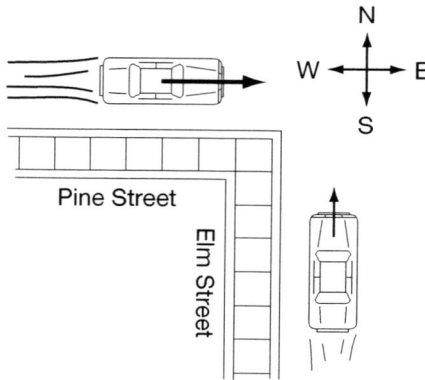

Quantities that do not have a direction are known as **scalars.** For example, temperature and mass are scalars. You cannot say "the temperature is 70° north." You cannot say, "My mass is 50 kg south," either. Therefore, these quantities are scalars.

Practice 1—Translation and Vectors

Directions: Circle the correct answer to each of the following questions.

1. Which branch of mechanics describes the cause of motion?

   (a) kinematics   (b) dynamics   (c) optics

2. What term describes the motion of an object from one place to another?

   (a) rotation   (b) flotation   (c) translation

3. What term involves both speed and direction of motion?

   (a) acceleration   (b) displacement   (c) velocity

4. What do a pair of cars, one of which is traveling 20 mph north, and the other of which is traveling 20 mph south, have in common? ____.

   (a) speed   (b) velocity   (c) position

5. Which of the following is a vector quantity?

   (a) speed   (b) temperature   (c) velocity

6. Which of the following is a scalar quantity?

   (a) mass   (b) velocity   (c) displacement

Calculating with Speed and Distance

If you know how fast an object is moving, and for how long it has been moving, you can figure out how far it will move. Likewise, if you know how fast an object is moving, you can calculate how much time it will take to go a certain distance. Both of these calculations depend on one equation.

$$d = v \times t$$

In this equation, d represents distance (the magnitude of displacement), v represents speed (the magnitude of velocity), and t represents time. By inserting any two of these quantities into the equation, you can figure out the third.

©1998, 2001 J. Weston Walch, Publisher

To solve a problem with an equation, first write down the equation you are going to use. Then, write in the quantities you already know. You should be left with one unknown quantity. Solve the equation for this one variable, and you will have your answer.

Look at these examples below.

1. A car traveled at 60 miles per hour (speed or v) for 3 hours (time or t). What is the distance it traveled?

   $d = v \times t$

   $d = 60$ miles/hr $\times$ 3 hr

   $d = 60$ miles/hr $\times$ 3 hr (cancel the "hr"'s)

   $d = 60$ miles $\times$ 3 $= 180$ miles

   The car traveled 180 miles.

2. Suppose you can paddle a canoe at 3 miles per hr (speed or v), and you want to go 12 miles (distance or d). How long will it take?

   $d = v \times t$

   12 miles $= 3$ miles/hr $\times t$ (Divide both sides of the equation by 3 miles/hr and cancel)

   $12/3$ hr $= 1 \times t$

   4 hours $= t$

   It will take 4 hours to go 12 miles.

3. Suppose you shoot an arrow at a target 180 meters (distance or d) away, and it takes 1.2 seconds (time or t) to reach the target. What was the average speed of the arrow?

   $d = v \times t$

   180 m $= v \times 1.2$ sec

   180 m/1.2 sec $= 150$ m/sec $= v$ (Divide both sides by 1.2 sec)

   150 m/s $= v$

   The arrow's average speed was 150 meters per second.

This equation assumes that the object was moving at the same speed for the whole time, or that you know what the average speed of the object was. In the next section, we will learn how to describe the motion of an object whose speed or velocity changes.

Practice 2—Calculating with Speed and Distance

Directions: Circle the correct answer to each of the following questions.
(*Hint:* Remember the equation $d = v \times t$.)

1. An ant crawls 5 inches per second. If it crawls for 15 seconds, how far will it go?

   (a) 15 inches   (b) 75 inches   (c) 150 inches

2. A car drives at 45 miles per hour. If the car travels for 3 hours at this speed, how far can it drive?

   (a) 15 miles   (b) 45 miles   (c) 135 miles

3. Miguel can run the 100-meter dash in 20 seconds. What is his average speed?

   (a) 5 meters per sec

   (b) 80 meters per sec

   (c) 120 meters per sec

4. Malika can ride a bicycle at an average speed of 30 kilometers per hour. How long will it take her to ride 120 km?

   (a) 4 hours   (b) 6 hours   (c) 8 hours

Acceleration

You are probably familiar with the terms *acceleration* and *deceleration*. They are everyday words used to refer to increases (acceleration) and decreases (deceleration) in the speed of objects. For example, when we say that a car accelerated to 60 miles an hour, we mean that the speed of the car increased to 60 miles/hour.

In the study of physics, the term acceleration has a more specialized meaning. **Acceleration** is how quickly an object changes its velocity. (Remember that velocity refers to both speed and direction.)

Acceleration can describe an object whose speed is increasing. For example, a car that goes from 0 mph to 60 mph in 5 seconds has a greater acceleration than a car that takes 15 seconds to reach 60 mph.

Acceleration can also describe an object whose speed is decreasing. For example, if a train is slowing down, or decelerating, we say it has a negative acceleration.

Acceleration can also describe an object whose speed does not change, but whose direction is changing. For example, an airplane that changes its course from north to northwest is accelerating. If the plane changes direction more quickly, its acceleration is greater.

Acceleration is how quickly the velocity is changing over time. Therefore, to describe the acceleration of an object, you need to tell how much the velocity changes in a certain amount of time. For example, the three sentences below describe the acceleration of objects.

> *Each second it fell, the rock fell 32 feet per second faster.*

> *The train slowed down by 15 miles per hour in one minute.*

> *The course of the airplane changed from north to northwest at a rate of 3° per second.*

You can calculate the acceleration of an object using the following equation.

$$a = v/t$$

Here, v is the change of velocity, a is acceleration, and t is time.

For example, if a car goes from 0 mph to 60 mph in 5 seconds, then its acceleration is:

$$a = v/t$$

$$a = 60 \text{ miles/hr/5 seconds}$$

$$a = 12 \text{ miles/hr/sec}$$

Read the answer "12 miles per hour, per second." This means that every second, the speed of the car changes by 12 miles per hour.

You can also use this equation to predict the speed of a moving object. For example, if the car continued with the same acceleration, how fast would it be going after 7 seconds?

$$a = v/t$$

$$12 \text{ miles/hr/sec} = v/7 \text{ sec}$$

Multiply both sides by 7 seconds.

$$84 \text{ miles/hr} = v$$

So, after 7 seconds, the car would be going 84 miles/hr.

This equation assumes that the acceleration is **constant,** or unchanging. In other words, the object continues to change its speed at the same rate. Obviously, this is not completely true in all cases. A car can usually accelerate from 0 to 20 mph much faster than it can accelerate from 60 to 80 mph. At low speeds, a car has a high acceleration, while at high speeds its acceleration decreases.

However, this equation is still useful if you use the *average* acceleration. Furthermore, some objects, such as falling objects, do exhibit constant acceleration, as you will see in the lesson on gravity later in this book.

Practice 3—Acceleration

Directions: Circle the correct answer to each of the following questions.

1. What term describes how quickly an object's velocity is changing?

    (a) speed

    (b) acceleration

    (c) rotation

2. Which of the following does NOT describe acceleration?

    (a) The car increased its speed by 5 mph every second.

    (b) The plane changed its direction from north to west in 3 minutes.

    (c) The man ran 8 miles each hour.

3. If a car increases its speed from 0 mph to 30 mph in 5 seconds, what is its acceleration?

    (a) 30 mph

    (b) 5 miles / hour / sec

    (c) 6 miles / hour / sec

4. If a plane accelerates at 30 miles / hour / sec, how fast will it be going after 6 seconds?

    (a) 5 miles / hour

    (b) 180 miles / hour

    (c) 300 miles / hour

Describing Rotational Motion

Rotational motion is the term for an object that spins without moving from one place to another. In this section, you will learn that many of the terms you learned for translational motion can be applied to rotational motion.

Every spinning object has an **axis of rotation.** This is the imaginary line that the object spins around. For example, Earth's axis of rotation runs straight through Earth from the North Pole to the South Pole; Earth spins around this line. See the diagrams on page 11 for the axis of rotation of Earth and two other objects.

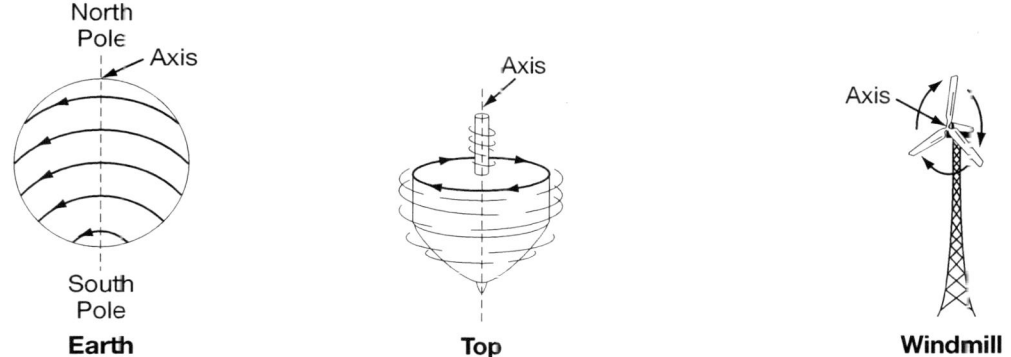

Earth **Top** **Windmill**

To describe how far an object moves in translational motion, we use *displacement* or *distance*. In rotational motion, the object does not move from one place to another. Instead, we describe the amount of rotation of the object using angles.

If you choose any point on the object and measure the angle it moves, we call this the **angular displacement** of the object. In the diagram below, the angular displacement is 60°. A half rotation is 180°. One complete rotation would be 360°.

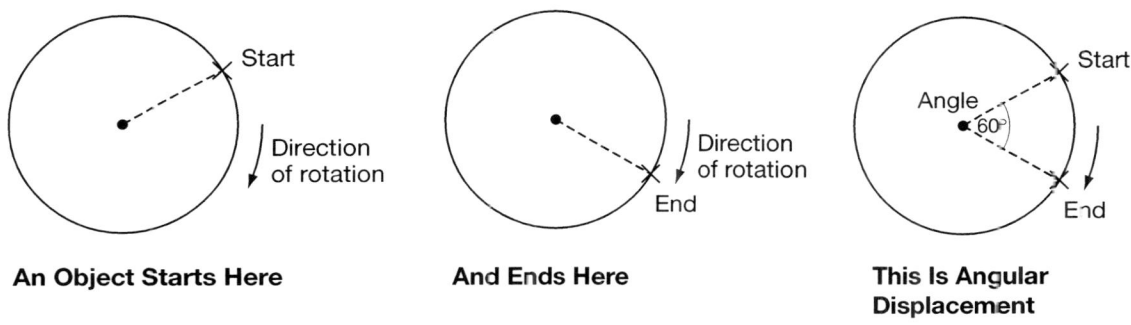

An Object Starts Here **And Ends Here** **This Is Angular Displacement**

If an object spins many times, it is inconvenient to talk about angular displacement in degrees. Instead, we talk about the number of **revolutions,** or complete turns, the object makes. For example, if an object spins 20 times, we say its angular displacement is 20 revolutions.

In translational motion, to describe how fast the object moves, we use velocity. Velocity is how far the object moves *per time*, such as 50 miles *per hour*, or 10 meters *per second*. Similarly, for rotational motion, **angular velocity** is how fast the object is spinning. Angular velocity is how much the object rotates per time. For example, we can say Earth spins 360° per day, or a top makes 10 revolutions per second.

A term you may be familiar with is **rpm.** This stands for **revolutions per minute.** For example, a 45 rpm record makes 45 complete turns per minute. A 33⅓ rpm record spins more slowly; it only makes 33⅓ revolutions each minute.

Your car probably has a dial labeled "rpm." This dial measures the number of rotations the engine's crankshaft makes every minute. If the rpm is too high, it means the engine is turning too fast. This can damage the car, because the faster the engine rotates, the more friction it creates. The same thing happens when you rub your hands together quickly. This friction creates heat that can ruin the engine.

Finally, we can talk about the **angular acceleration** of an object. Angular acceleration is how much the spinning of an object is speeding up or slowing down. In other words, it is how quickly its angular velocity is changing.

For example, when your car accelerates, the wheels have an angular acceleration. When you drive at a constant speed, the wheels have no angular acceleration. When you spin a top and it begins to slow down, the top has a negative angular acceleration; its angular velocity is decreasing.

Scientists have discovered that Earth actually has a small negative angular acceleration. It is gradually slowing down, just like a top! What does this mean? What will happen as a result of this slowing down?

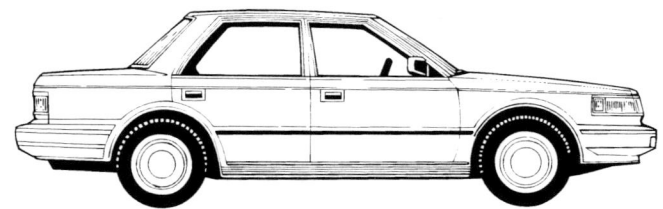

Practice 4—Describing Rotational Motion

Directions: Circle the correct answer to each of the following questions.

1. Which of the following is an example of rotational motion?

    (a) a car driving up a hill

    (b) a car driving around a turn

    (c) the spinning of a car's wheels

2. What is the imaginary line around which an object spins?

    (a) axis of rotation

    (b) point of acceleration

    (c) axis of translation

3. What term describes how quickly an object is spinning?

    (a) angular displacement

    (b) angular velocity

    (c) angular acceleration

4. How fast is an object spinning that spins at 50 rpm?

    (a) 50 times every second

    (b) 50 times every minute

    (c) 50 times every hour

5. What is the angular displacement of an object that spins 20 times?

    (a) 20 revolutions

    (b) 20 degrees

    (c) 20 miles per hour

■ LESSON MASTERY TEST

Directions: Decide whether each sentence below describes translational (**T**) or rotational (**R**) motion. Write the correct letter in each blank.

_____ 1. Earth turns on its axis.

_____ 2. An ant crawls across the floor.

_____ 3. A bird flies through the air.

_____ 4. A cassette's reels turn in a tape player.

Directions: Decide whether each statement below describes displacement (**D**), speed (**S**), velocity (**V**), or acceleration (**A**). Write the correct letter in each blank.

_____ 5. A car is driving southwest at 45 miles per hour.

_____ 6. The car drove north for 50 miles.

_____ 7. The car slowed down as it approached the red light.

_____ 8. The car was going faster than the speed limit.

Directions: Circle the correct answer to each of the following questions.

9. What is the imaginary line around which an object spins?

   (a) axis of rotation

   (b) angular acceleration

   (c) axis of translation

10. Which of the following quantities has direction and magnitude?

    (a) scalar

    (b) vector

    (c) temperature

(continued on next page)

11. A car is traveling at 80 kilometers per hour. If it continues at this speed for 2.5 hours, how far will it go?

   (a) 80 kilometers

   (b) 100 kilometers

   (c) 200 kilometers

12. Rosa walked at a rate of 50 meters per minute. How long will it take her to walk 1 kilometer (1,000 meters)?

   (a) 10 minutes

   (b) 20 minutes

   (c) 50 minutes

13. What is the acceleration of a car that increases in speed from 0 mph to 30 mph in 5 seconds?

   (a) 5 miles / hour / sec

   (b) 6 miles / hour / sec

   (c) 30 miles / hour / sec

©1998, 2001 J. Weston Walch, Publisher

■ Lesson 2—Dynamics

Goals: To understand Newton's Laws of Motion; to use those laws to explain why things move the way they do

Newton's First Law and Inertia

The ancient Greek scientists believed that the natural state of all objects was to be still, or not moving. For example, a ball will not roll unless it is pushed. If you roll the ball across a carpet, it slows down and stops. It won't start moving again until somebody pushes it.

However, in the 1600's, the English scientist **Isaac Newton** realized that there is actually a force making the ball stop—friction. If you roll the ball on a smooth floor, it will roll much farther before stopping than it would on a carpeted floor or other rough surface. If you could completely eliminate friction, the ball could roll forever. This led to Newton's first law of motion, the principle known as **inertia.**

Look at the two diagrams below. If the woman does not push the cart, will it move by itself? Of course not. What will happen to the baseball if the batter does not hit it? It will continue moving until some other force makes it stop.

These examples show that the natural state of an object is to continue the same motion. An object that is not moving will continue to stay still. An object that is moving will continue moving. A force, some kind of push or pull, is needed to change the object's motion.

Putting this in terms you learned in Lesson 1, the velocity of an object will not change unless there is some force. An object will move in the same direction, at the same speed, unless a force causes some acceleration.

This principle is described in **Newton's First Law,** which states:

> **Unless a force acts on it, an object at rest will stay at rest, and an object in motion will continue to move in a straight line at the same speed.**

This is known as the **principle of inertia.**

On Earth, there is almost always some force acting on an object. Things rolling on the ground, or moving through the air, experience friction. And, gravity is a force that tries to pull all things down to the ground. That's why a baseball doesn't actually move in a straight line—it always slows down and falls to the ground.

If there is always friction and gravity, then why is the principle of inertia useful? Because it helps explain many cases where the outside forces aren't so important. For example, if you are driving in a car at 30 mph, and you flip a coin, why doesn't the coin fly to the back of the car? Its inertia causes it to continue moving along with the car.

On the other hand, if the car stops suddenly, the passengers will be thrown forward. Why? Because their inertia causes them to keep moving at 30 mph, even though the car stops. Some force, such as a seatbelt, is needed to change their velocity.

When a car goes around a curve, the passengers are pushed to the outside. Again, this is inertia. If the car is going north at first, and then turns, the inertia of the passengers makes them continue to move north, even though the car is turning. Again, the force of the seatbelt, or the side of the car, is needed to make them turn.

You may have seen a magic trick where a magician pulls a tablecloth out from under plates and glasses. This is not magic, but inertia. The plates will stay still unless a force acts on them. If there is very little friction between the tablecloth and the plates, then the plates won't move.

In outer space, when a spaceship turns off its engines, will the spaceship stop?

Practice 1—Newton's First Law and Inertia

Directions: Circle the correct answer to each of the following questions.

1. What term describes the tendency of a stationary object to remain at rest?

    (a) acceleration   (b) inertia   (c) kinematics

2. What happens to a moving object if there are no forces acting on it?

    (a) It will not move.

    (b) It will accelerate.

    (c) It will continue moving the same way.

3. What causes a ball rolling on the ground to eventually come to rest?

    (a) inertia   (b) friction   (c) mass

Directions: Decide whether each statement that follows is true (**T**) or false (**F**). Write the correct letter in each blank.

_____ 4. When the driver of a car puts on the brakes, the passengers move forward because the car has a lot of inertia.

_____ 5. A rocket in outer space will gradually slow down due to inertia.

_____ 6. If there are no forces acting on an object, its velocity will not change.

Newton's Second Law

All objects have inertia. It takes some force in order to change their motion. But do all objects have the same amount of inertia? In other words, does it require more force to move some objects than it does to move others?

It certainly does. For example, it takes more force to throw a bowling ball than a golf ball. The difference is that the bowling ball has more mass. Objects with more mass have more inertia; they are harder to accelerate.

Newton's Second Law of Motion states:

> **A force causes an object to accelerate, and the more mass an object has, the more force you need to get the same acceleration.**

It is important to remember that a force causes an acceleration. This means that as long as a force is acting on the object, it will continue to change its speed. The longer you press on the gas pedal in your car, the more its speed will increase.

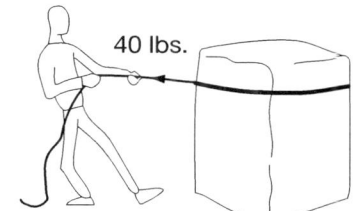

Figure 1

It is also important to recognize that forces can cancel each other out. For example, suppose you and a friend pull on opposite sides of a rope with equal amounts of force. The rope won't move, because your forces cancel each other out. Therefore, we often talk about the **net force** on an object. This is the total force, taking into account forces that cancel each other out. In the example of the rope, there is no net force, because the two forces being applied cancel each other out.

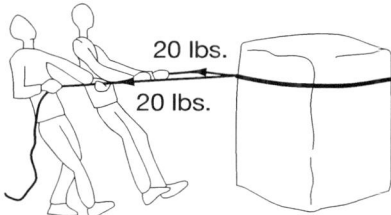

Figure 2

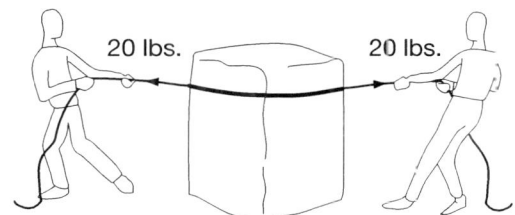

Figure 3

The figures on the right show different examples of net force. In Figure 1, the net force is 40 pounds because there is only one force. In Figure 2, the net force is again 40 pounds because the two forces help each other. In Figure 3, the net force is 0 because two equal forces, pulling in opposite directions, completely cancel each other out. In Figure 4, the net force is 20 pounds toward the left, because the force to the right decreases the net force by 10 pounds.

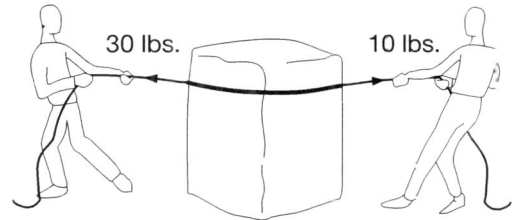

Figure 4

Net force also explains why a vehicle cannot go faster and faster with no limit. This is because as the vehicle moves faster, the wind resistance on the vehicle increases. Eventually, the wind resistance is so great that it cancels out the force the engine provides. No net force means the vehicle cannot accelerate anymore, so its speed will not increase.

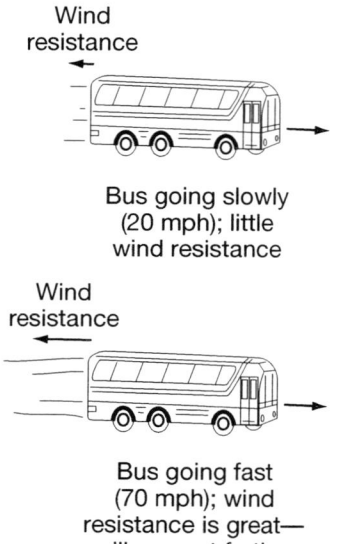

Wind resistance

Bus going slowly (20 mph); little wind resistance

Wind resistance

Bus going fast (70 mph); wind resistance is great— will prevent further acceleration

Suppose you apply a force (pull) of 25 pounds on a wagon, and it accelerates 1 mph per second. If you apply half as much force, it will only accelerate ½ mph per second. If you apply twice as much force, it will accelerate 2 mph per second.

However, suppose you put some bricks in the back of the wagon, so that it has twice as much mass. Then you would only get half as much acceleration. So, **changing the force or changing the mass will change the amount of acceleration.**

In Real Life

Sometimes car designers want to increase the acceleration of a car. Car designers can do this by giving the car a smoother and more aerodynamic shape which will decrease the air resistance of the car but not change the amount of force the engine provides. By doing so, the air resistance would not cancel out as much of the force of the engine, and the net force would be greater. The designer could also decrease the mass of the car. Then, the same force would provide more acceleration.

This is summarized in the following equation.

$$F = m \times a$$

Here F is force, m is mass, and a is acceleration. This equation allows you to predict the acceleration of an object, if you know its force and its mass. Likewise, you can predict the force needed to get a certain acceleration. Look at the examples below.

1. Suppose your car weighs 1,000 kg, and you want it to accelerate at 5 m/sec/sec. How much force must the engine provide?

    $$F = m \times a$$

    $$F = 1,000 \text{ kg} \times 5 \text{ m/sec/sec} = 5,000 \text{ kg} \times \text{m/sec/sec}$$

    $$F = 5,000 \text{ N}$$

    The engine must provide 5,000 N of force. Remember, in the metric system, force is measured in Newtons
    (1 N = 1 kg × m/sec/sec).

2. Suppose you put an even more powerful engine in the same car, so it can now provide 8,000 N of force. What will its acceleration be now?

    $$F = m \times a$$

    $$8,000 \text{ N} = 1,000 \text{ kg} \times a$$

    $$8,000 \text{ N}/1,000 \text{ kg} = a \text{ (Divide both sides by 1,000 kg.)}$$

    $$8 \text{ m/sec/sec} = a$$

    It will now accelerate 8 m/sec/sec.

Practice 2—Newton's Second Law

Directions: Circle the answer that correctly completes each of the following statements.

1. Objects with more mass have ____ inertia than objects with less mass.

   (a) more   (b) less   (c) the same

2. If you apply the same force to two objects, the one with more mass will accelerate ____.

   (a) more   (b) less   (c) the same amount

3. Applying a constant force gives an object constant ____.

   (a) mass   (b) velocity   (c) acceleration

Directions: Circle the correct answer to each of the following questions.

4. What is the net force in the diagram to the right?

   (a) 0 N

   (b) 20 N to the left

   (c) 40 N to the left

   (d) 40 N to the right

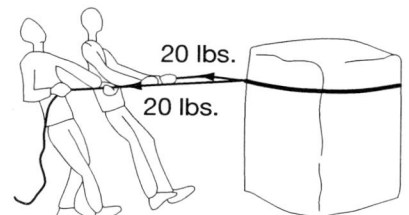

5. What is the net force in the diagram to the right?

   (a) 0 N

   (b) 10 N to the left

   (c) 20 N to the left

   (d) 30 N to the left

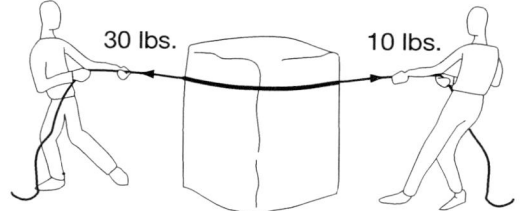

6. How much force is required to accelerate a bicycle and rider, with combined mass of 80 kg, at a rate of 3 m / sec / sec?

   (a) 30 N   (b) 80 N   (c) 240 N

The Direction of Forces

So far, we have talked only about *how much* acceleration a force causes. But, force is a vector, meaning it has a strength *and* a direction. So, how does the direction of a force affect the motion?

Imagine your friend is gliding on ice skates. If you push him from the back, you are applying a force forward and his speed will *increase*. If you push him from the front, you are applying a force backward and he will slow down. If you push him from the side, he will turn. In other words, his *direction* of motion will change, even if his speed does not. Remember, acceleration means a change in velocity, so it can cause either a change in speed or a change in direction.

In fact, if you apply the force **perpendicular** (see the diagram below) to the motion, only the direction will change, and the skater will turn in the direction the force was applied. The speed will be unaffected, while the direction changes.

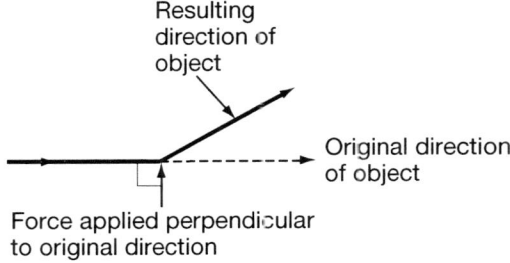

Resulting direction of object

Original direction of object

Force applied perpendicular to original direction

As an example of this, look at the diagram of the boy swinging a rock. The rock's direction is constantly changing. This means it is accelerating. What force is making it go in a circle?

Inward force

The string pulling inward on the rock forces it to go in a circle. If the string broke, the rock would fly off in a straight line, because of inertia. But the force of the string, which is perpendicular to the motion of the rock, causes the rock's direction to continually change, going in a circle. An inward perpendicular force like this, which causes an object to move in a circle, is called **centripetal force.** Later in this book, you will see that gravity provides a similar centripetal force. This is what makes the moon go around Earth, and Earth go around the sun.

Tip!

Remember: Applying a force in the same direction as the motion of an object causes the speed to increase. Applying a force in the opposite direction causes the speed to decrease. If you apply a force at a right angle, only the direction will change. A force applied in some other direction affects both the speed and direction.

Practice 3—The Direction of Forces

Directions: Circle the answer that correctly completes each of the following statements.

1. Force is _____.

    (a) a vector

    (b) an acceleration

    (c) a direction

2. In order to slow down an object, you must apply a force _____ the motion.

    (a) in the same direction as

    (b) opposite to

    (c) perpendicular to

3. In the diagram to the right, the force applied will cause the object to _____.

    (a) speed up

    (b) slow down

    (c) change direction

    (d) all of the above

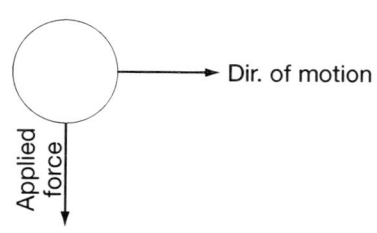

Dir. of motion

Applied force

4. A centripetal force causes an object to _____.

    (a) move in a straight line

    (b) move in a circle

    (c) slow down

Newton's Third Law

When you hit a nail with a hammer, the force the hammer exerts pushes the nail into the board. But what causes the hammer to stop moving? The nail exerts a force back on the hammer, causing it to stop.

If you kick a wall with your bare foot, your foot will hurt. This is because when you exert a force on the wall, the wall exerts a force back on you.

These are both examples of **Newton's Third Law of Motion.** It states:

> **For any force, there is an equal** *reaction* **force, in the opposite direction.**

The hammer pushes the nail, and the nail pushes back equally hard on the hammer. You kick the wall, and the wall pushes back just as hard.

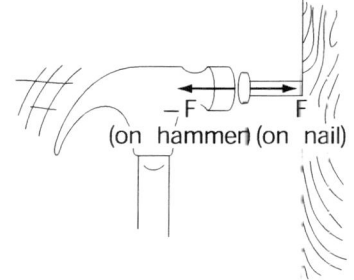

The hammer exerts a force (F) on the nail; the nail exerts a force (−F) on the hammer.

−F represents the reaction force.

Think About It

Identify the reaction force for each of the following examples of force: (a) The blades of a helicopter push the air downward. (b) A weight-lifter pushes up on a barbell. (c) A car slams into a telephone pole.

It is important to remember that **the reaction force is always equal to the original force.** The harder you kick the wall, the harder it pushes back on your foot. Imagine standing on roller skates and throwing a basketball. You will roll backward because of the reaction force. The harder you push the basketball, the more you will roll backward. This is because you feel a greater reaction force, which, by Newton's Second Law, means a greater acceleration.

It is important to remember the mass of the objects involved in Newton's Third Law. For instance, if gravity pulls a rock toward the ground, then Newton's Third Law says that the rock is also pulling Earth toward it. So, why doesn't Earth move each time we drop a rock? Because the mass of Earth is so great, it has a tiny acceleration that we could never even measure.

Think About It

Suppose a small child and a large adult are on roller skates. If they both throw a basketball at the same speed, who will roll backward faster?

Similarly, when a gun is fired, the bullet pushes back on the gun with a force equal to the gun pushing on the bullet. So, why doesn't the gun kick back as fast as the bullet goes forward? Because the mass of the gun is greater, so its acceleration is less.

Practice 4—Newton's Third Law

Directions: Circle the correct answer to each of the following questions.

1. When a hammer hits a nail, with what force does the nail push back?

   (a) less   (b) more   (c) equal

2. For any force, there is an equal reaction force in what direction?

   (a) in the same direction

   (b) in the opposite direction

   (c) perpendicular

3. Which of the following is NOT an example of Newton's Third Law?

   (a) The blades of a helicopter push the air down so the air pushes the helicopter up.

   (b) You push a light switch down so the lights turn off.

   (c) A person pushes a heavy object and falls backward.

Directions: Decide whether each statement that follows is true (**T**) or false (**F**). Write the correct letter in each blank.

_____ 4. When a gun is fired, the force the gun exerts on the bullet is equal to the force of the bullet pushing back on the gun.

_____ 5. When a gun is fired, the bullet accelerates more than the gun, because the mass of the bullet is less than the mass of the gun.

_____ 6. When an apple falls, gravity pulls the apple toward the Earth, but the apple exerts no force on the Earth.

Rotational Dynamics

In translational motion, you learned that objects have inertia. An object that is moving will continue moving, and an object at rest will stay at rest, unless a force acts on it.

Spinning objects also have inertia. We call it **rotational inertia.** A stationary object requires a push to start it spinning. An object will continue spinning at the same rate until a force speeds it up or slows it down.

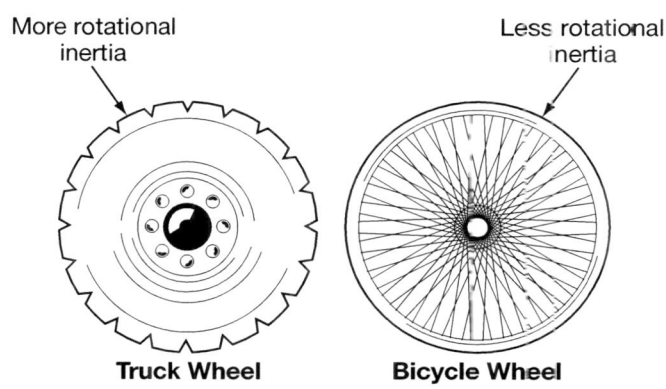

More rotational inertia

Less rotational inertia

Truck Wheel **Bicycle Wheel**

Translational inertia depends only on the mass of the object. The more mass it has, the more difficult it is to change its velocity. For spinning objects, mass also affects the rotational inertia. For example, the heavy truck wheel above is more difficult to turn than the bicycle wheel.

In addition, the distribution or shape of the mass is important for rotational inertia. In the diagram below, it is easier to rotate the stick with the nail through the center. It has less rotational inertia. Similarly, it is easier to rotate the hammer when it is held near the head than when it is held in the center.

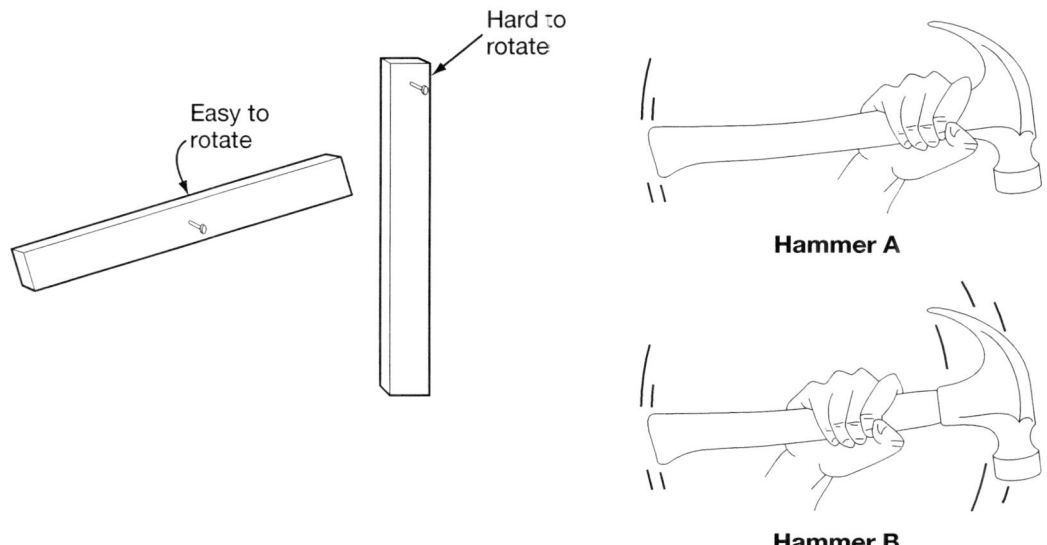

Hard to rotate

Easy to rotate

Hammer A

Hammer B

So, rotational inertia depends on where the axis of rotation is. If most of the mass is near the axis of rotation, as for Hammer A on page 27, then it will have less rotational inertia. It will be easier to rotate.

Physicists can use a formula that takes into account the distribution of mass in an object to determine the rotational inertia of that object spinning around a specific axis of rotation.

In translational motion, force is simply a push or pull on an object, which causes an acceleration. A force of the same magnitude always has the same effect on the velocity of a particular object.

In rotational motion, however, the effect of a force depends on where it is applied. **The farther the force is from the axis of rotation, the more effect it will have.** For example, a force at point A on the bicycle wheel below will cause the wheel to spin faster than the same force at point B. Similarly, it will be easier to push the board by pushing it at point C rather than at point D.

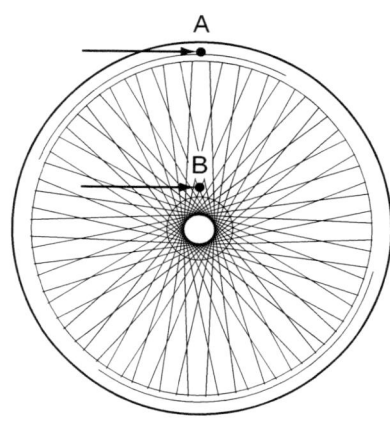

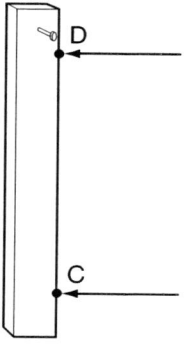

The direction of the force also determines its effect on rotational motion. For example, pushing the board vertically, as in the first diagram on the right, will have no effect on rotation. You must push the board horizontally, as shown on the right. **Any force that points directly toward the axis of rotation will have no effect.** The force must point perpendicular to the axis of rotation, as in the second diagram.

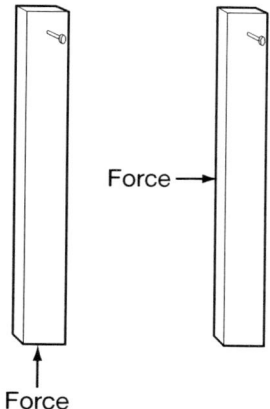

For rotational motion, the equivalent of force is called **torque.** Torque takes into account the strength of the force, as well as where it is applied and in which direction. **The more torque a force causes, the greater angular acceleration it will create.**

In cases where the force is perpendicular to the axis of rotation, there is a simple formula to determine the torque.

$T = F \times d$

Here T is torque, F is force, and d is the distance between the force and the axis of rotation. The greater the force, the more torque it creates. Likewise, increasing the distance from the axis of rotation creates more torque.

Which force will create more torque in the diagrams below?

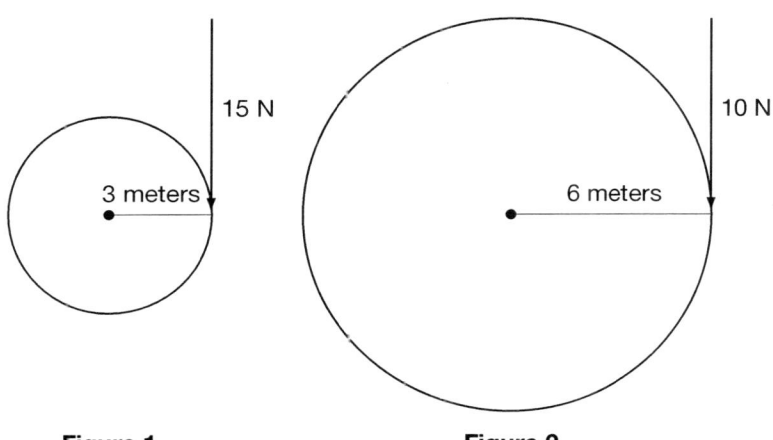

Figure 1   **Figure 2**

In Figure 1:

$T = F \times d$

$T = 15 \text{ N} \times 3 \text{ m} = 45 \text{ N-m}$
        (N-m is a Newton meter, a unit used in measuring torque.)

In Figure 2:

$T = F \times d$

$T = 10 \text{ N} \times 6 \text{ m} = 60 \text{ N-m}$

So, in Figure 2, even though the force is smaller, it creates a greater torque. If the objects have equal rotational inertia, then the object in Figure 2 will rotate faster.

Translational and rotational motion have many similar concepts. For example, in translation you learned about displacement, velocity, and acceleration. In rotation, these are called angular displacement, angular velocity, and angular acceleration. Similarly, in translational dynamics, you learned about inertia and forces. In rotational dynamics, the corresponding terms are rotational inertia and torque. The names are different to distinguish between translation and rotation, but the ideas behind them are similar.

Practice 5—Rotational Dynamics

Directions: Circle the correct answer to each of the following questions.

1. What is the resistance of a rotating body to changes in angular velocity called?

   (a) axis of rotation

   (b) rotational inertia

   (c) torque

2. Which of the following objects will be easier to rotate?

   (a) the stick with the nail in the center

   (b) the stick with the nail at the end

   (c) no difference

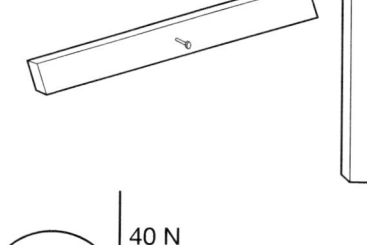

3. What is the torque created by the following force?

   (a) 40 N-m

   (b) 80 N-m

   (c) 240 N-m

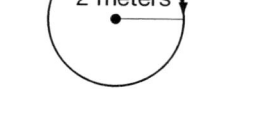

Directions: Decide whether each statement that follows is true (**T**) or false (**F**). Write the correct letter in each blank.

_____ 4. A 5 kg object with most of its mass near the axis of rotation will have less rotational inertia than a 5 kg object with its mass far from the axis of rotation.

_____ 5. A larger force always creates a larger torque.

_____ 6. Increasing the distance from the axis of rotation creates more torque.

_____ 7. Spinning objects resist changes in their angular velocity.

■ LESSON MASTERY TEST

Directions: Decide whether each of the following statements is an example of Newton's First Law (**A**), Newton's Second Law (**B**), or Newton's Third Law (**C**). Write the correct letter in each blank.

_____ 1. The car was able to stop, but the large truck behind it could not stop before hitting the car.

_____ 2. The soldier was injured when the cannon recoiled after he had shot t.

_____ 3. If it weren't for the gravity of the sun, Earth would fly off in a straight line into space.

_____ 4. To make a car faster, an engineer designs a smaller and lighter body, but keeps the power of the engine the same.

_____ 5. A squid draws water into its body opening, and then forces the water out at high speed, sending itself traveling in the opposite direction.

Directions: Circle the answer that correctly completes each of the following statements.

6. If a force is applied at the outside edge of a bicycle wheel, it will start the wheel spinning _____ the same force applied closer to the axle of the wheel.

   (a) at the same speed as

   (b) faster than

   (c) slower than

7. The effect of a force that rotates a body is called _____.

   (a) angular velocity

   (b) torque

   (c) axis of rotation

(continued on next page)

8. An inward force that makes an object move in a circle is called a ____.

   (a) reaction force

   (b) centripetal force

   (c) angular force

9. The reaction force is ____ the original force.

   (a) in the same direction as

   (b) in the opposite direction as

   (c) perpendicular to

10. Applying a constant force gives an object constant ____.

   (a) mass

   (b) velocity

   (c) acceleration

Directions: Circle the correct answer to the following question.

11. A wagon full of bricks has a mass of 100 kg. How much force is required to accelerate it at 3 meters / sec / sec?

   (a) 30 N

   (b) 100 N

   (c) 300 N

■ Lesson 3—Gravity

Goal: To understand the effects of gravity, both on Earth and in the universe

The Law of Falling Objects

In the early 1600's, the Italian scientist **Galileo Galilei** began to study the motion of falling objects. Before his time, few scientists did experiments. They simply thought about something, and came up with the most reasonable answer. However, many of their answers were incorrect. Galileo made careful measurements of objects falling or rolling down a ramp, and discovered the "Law of Falling Objects."

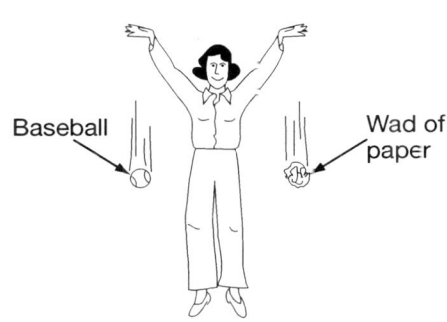

Baseball

Wad of paper

The first question he asked was whether heavy objects fall faster than light objects. He found, to the surprise of many people, that a small brick and a large brick dropped from a tall building hit the ground at the same time. He concluded that gravity does not make heavier objects fall faster.

Of course, you know that a brick falls faster than a feather. However, that is due to air resistance, rather than gravity. Air resistance (friction with the air) slows the feather more than it slows the brick. However, if there was no air, the feather and brick would fall at the same rate. And objects with little air resistance, such as a large brick and a small brick, actually do fall at nearly the same rate.

People have tested Galileo's theory by removing the air from a container. When a feather and a brick are dropped in such a container, they really do fall at the same rate! They have no air resistance.

The other thing Galileo observed was that a falling object does not fall the same distance each second. In the first second, it falls 5 m; after two seconds, it has fallen 20 m; after three seconds, it has fallen 45 m. There seems to be no pattern, until you observe the speed of the object. After the first second, it is moving at 10 m/s. After two seconds, it is moving at 20 m/s. After three seconds, 30 m/s, and so on. The speed is increasing, but it is increasing at a constant rate. **A falling object changes speed, but has a constant acceleration.** It speeds up by about 10 m/s each second.

Distance and Speed of a Falling Object		
time	distance fallen	speed
1 sec	5 m	10 m/sec
2 sec	20 m	20 m/sec
3 sec	45 m	30 m/sec
4 sec	80 m	40 m/sec

Galileo summarized his findings in the **Law of Falling Objects.**

All objects, regardless of weight, fall with the same constant acceleration.

On Earth, the acceleration of a falling object is about 10 m/sec/sec, or 32 feet/sec/sec. This means that each second, an object falls 10 m/sec or 32 feet/sec faster.

Because falling objects have a constant acceleration, you can use the equation you learned in the first lesson to predict its speed. Look at the examples below.

1. How fast will a brick move if it falls for 6 seconds?

    $a = v/t$

    $v = t \times a$

    so

    $v = 6 \text{ sec} \times 10 \text{ m/sec/sec}$

    $v = 60 \text{ m/sec}$

The brick will move at 60 m/sec.

However, because the speed is changing, we cannot use the equation $d = v \times t$ to find the distance it falls. This requires a new equation, based on acceleration.

$$d = 1/2\ (a) \times (t^2)$$

Remember, t^2 is the same as $t \times t$.

2. How far will a brick fall after 5 seconds?

$$d = 1/2\ (a) \times (t^2)$$

$$d = 1/2\ (10\ \text{m/sec/sec}) \times (5\ \text{sec})^2$$

$$d = 1/2\ (10\ \text{m/sec/sec}) \times 25\ \text{sec}^2$$

$$d = 5\ \text{m/sec/sec} \times 25\ \text{sec}^2$$

$$d = 125\ \text{m}$$

It will fall 125 meters.

Galileo's discovery of the law of falling objects opened the modern era of scientific study. He used careful measurements to determine that all objects fall with the same acceleration. Then, he described equations that could be used to predict the distance and speed of a falling object. This was the beginning of the use of mathematics to describe nature.

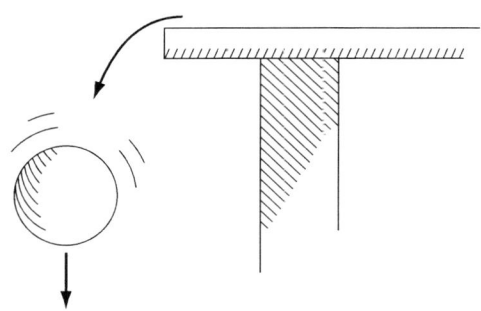

Tip!

To understand how objects fall, you must remember that they do not fall with constant speed, but with *constant acceleration*. The speed is continually increasing, but the rate that the speed is increasing is constant. Every second, the speed increases by 10 m/sec.

Practice 1—The Law of Falling Objects

Directions: Circle the answer that correctly completes each of the following statements.

1. Galileo determined that all objects fall with constant ____.

   (a) position

   (b) speed

   (c) acceleration

2. Each second, a falling object increases its ____ by about 10 meters per second.

   (a) distance fallen

   (b) speed

   (c) acceleration

3. Ignoring air resistance, a heavy object falls ____ a light object.

   (a) faster than

   (b) slower than

   (c) at the same speed as

4. A falling feather is slowed down due to ____.

   (a) air resistance

   (b) gravity

   (c) weight

Directions: Circle the correct answer to each of the following questions.

5. If a brick falls for 9 seconds, what will its velocity be? (Remember $a = v / t$. The acceleration of an object falling on Earth is 10 m/sec/sec.)

   (a) 9 m / sec   (b) 19 m / sec   (c) 90 m / sec

6. How far will a bowling ball fall in 6 seconds?
   (Remember, for accelerating objects, $d = 1/2\, a \times t^2$.)

   (a) 100 meters   (b) 180 meters   (c) 360 meters

Newton's Law of Universal Gravitation

You just learned about Galileo's law of falling objects. He discovered that on Earth, all objects fall with the same constant acceleration. However, he did not explain *why* this is true.

Around 1650, after formulating his three laws of motion, Isaac Newton began to think about why the moon moves around Earth in a circle. According to the first law of motion, if there were no force on the moon, it would travel in a straight line. Some force must be acting to make it change direction.

Remember from Lesson 2 that a **centripetal** or inward force is necessary to make something move in a circle. Therefore, there must be a force pulling the moon toward Earth, just like a string pulls on a rock swung overhead.

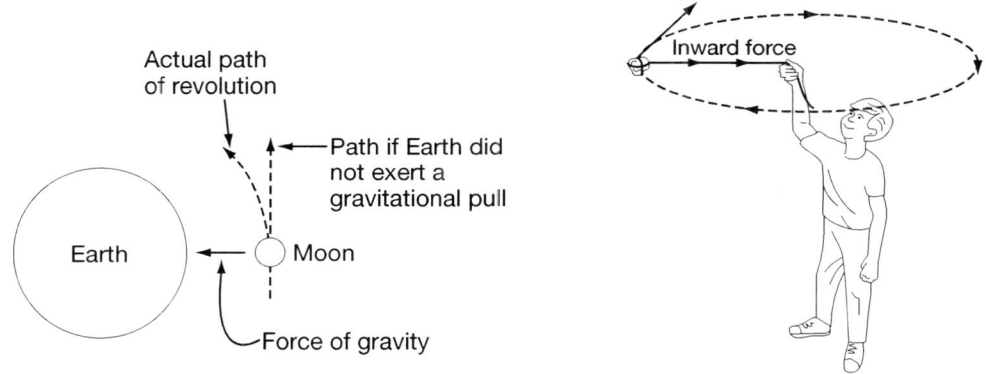

When an apple falls from a tree, there is also a force pulling it to Earth. Isaac Newton realized that these are both the same type of force. Gravity causes both the moon and the apple to fall toward Earth.

Furthermore, Earth is traveling around the sun in a circle. This means that the sun must have gravity also. Newton realized that, in fact, *all* objects have gravity. The sun, the moon, a mountain, a bowling ball, and even people, all exert gravitational pull on other objects. However, the strength of the pull depends on the mass of the objects and how far apart they are. This is known as **Newton's Law of Universal Gravitation.** It is called "Universal," because it says that the same force, gravity, affects every object, everywhere in the universe.

The force of gravity between you and an apple is very small, because your mass and the apple's mass are very small. However, the force of gravity between you and Earth is larger, because Earth has so much mass. If Earth had twice as much mass as it does now, the force of its gravity would be twice as much. This means you would weigh twice as much. On the moon, your weight would be less, because the moon has less mass than Earth.

Furthermore, the force of gravity depends on how far apart two objects are. Standing on the surface of Earth, you are about 6,000 kilometers from the center of Earth. If you were to go twice as far from the center of Earth, the force of gravity (your weight) would be ¼ as much. If you went three times as far away, it would be ⅑ as much.

So, all objects are exerting gravitational pull on each other. However, we notice only the strongest forces. The force between objects with small masses, or objects that are very far apart, are too small to be observed. Your desk is pulling on you, but you don't notice it. Jupiter is also pulling on you, but it is too far away to exert a powerful force like Earth's pull.

The force of gravity also depends on the mass of the object being pulled. That is why a bowling ball feels heavier than a golf ball. The bowling ball has more mass. In fact, an object with twice as much mass feels twice as much gravitational pull. That is why it is easy to think (mistakenly) of weight and mass as the same. On Earth, mass and weight are proportional.

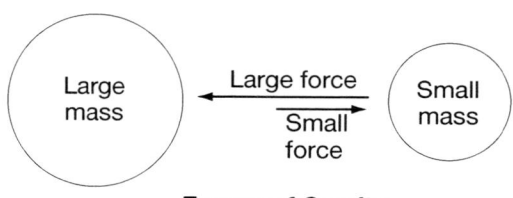

Forces of Gravity

If there is a stronger gravitational pull on a bowling ball than on a golf ball, then why did Galileo say they fall with the same acceleration? Because the bowling ball has more mass than the golf ball, it also has more inertia. Objects with more mass feel a greater gravitational pull, but they also have more inertia to resist movement. These two factors cancel each other out, making all objects fall at the same rate.

Practice 2—Newton's Law of Universal Gravitation

Directions: Circle the answer that correctly completes each of the following statements.

1. If Earth had twice as much mass, your weight would be ____.

   (a) the same

   (b) twice as much

   (c) half as much

2. The force of gravity on a 2 kg object is ____ as on a 1 kg object.

   (a) the same

   (b) twice as much

   (c) half as much

3. The acceleration due to gravity of a 2 kg object is ____ as that of a 1 kg object.

   (a) the same

   (b) twice as much

   (c) half as much

4. If the moon and earth were twice as far apart, the force of gravity between them would be ____.

   (a) twice as strong

   (b) half as strong

   (c) one-fourth as strong

Directions: Decide whether each statement that follows is true (**T**) or false (**F**). Write the correct letter in each blank.

____ 5. The force that makes the moon orbit Earth is different from the force that makes a brick fall.

____ 6. The force of gravity does NOT depend on how far apart the objects are.

____ 7. All objects exert a gravitational force on other objects.

Orbits of Planets and Satellites

To understand how Earth can pull the moon without the moon falling into Earth, consider this example. If you drop a ball, it falls straight down. If you throw it gently, it follows a curved path and hits the ground. If you throw it faster, it travels farther before hitting the ground.

Now, suppose you could throw the ball fast enough so that the curved path it makes, as it falls, exactly matches the curve of Earth. It would continue to fall, but would never reach Earth. In this case, it would now be in orbit around Earth.

Of course, you could never actually do this. For one thing, you would need to throw the ball at 17,500 miles per hour. Even if you had a strong enough arm to do this, the ball would burn up due to friction with the air.

However, this is exactly what is happening when the moon orbits Earth. The moon is falling toward Earth, but moving around it at the same time. It falls just enough each second to turn its motion into a continuous path around Earth. If its horizontal motion slowed down too much, it would fall into Earth. If it sped up too much, it would fly away from Earth.

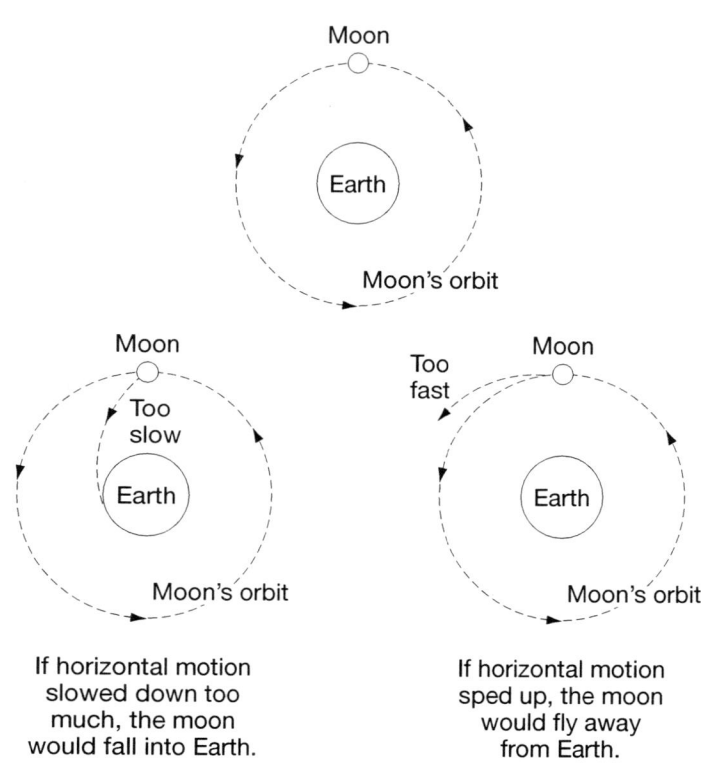

If horizontal motion slowed down too much, the moon would fall into Earth.

If horizontal motion sped up, the moon would fly away from Earth.

This same idea also keeps satellites and the Space Shuttle orbiting around Earth. Scientists use rockets to launch satellites above Earth's atmosphere (where there is no air resistance). They launch the satellites vertically to get them far up into space. Then they tilt the satellites and give them enough horizontal speed so that Earth curves away from them as they fall. The force of gravity is continuously pulling the satellites toward Earth, but the satellites' horizontal speed keeps them going in circles.

To reach this balance between inertia and gravity, a satellite must be moving at a precise speed. Since the pull of gravity decreases with distance, the velocity needed to keep a satellite in orbit changes with altitude. Engineers can calculate how high a satellite must be in order to orbit at a certain speed. For example, satellites close to Earth complete an orbit about once every 90 minutes. Other satellites orbit at the same speed as Earth's rotation—once every 24 hours. This allows them to stay directly above one point on Earth. These are known as **geosynchronous satellites.** They must orbit very far from Earth's surface, at an altitude of about 22,000 miles.

In order to understand orbits, just remember that an orbiting object is always being pulled toward Earth. However, it is also moving horizontally, so its falling path matches the curve of its orbit. It's as if the moon is falling, but keeps missing Earth as it falls.

Most planets and satellites do not orbit in perfect circles. They move in an oval shape, known as an **ellipse.** The sun is not at the center of the ellipse. So, the planets are continually moving closer to and farther away from the sun. As they get closer, they speed up, and as they move farther away, they slow down.

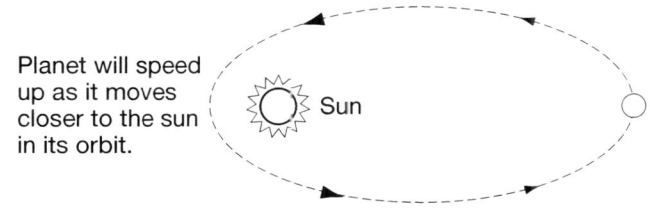

Planet will speed up as it moves closer to the sun in its orbit.

Sun

Think About It

You may have seen pictures of people in the Space Shuttle, floating as if there were no gravity. Is there really no gravity when they are in orbit?

If gravity is pulling on the astronauts, why do they appear weightless? The answer has to do with freefall. Imagine a person standing on a scale in an elevator. When the elevator is at rest, the person's weight is 150 pounds. When the elevator begins to go down, it drops out from under him. Gravity is at work. His weight will be slightly less—about 120 pounds. What if the cable breaks? The elevator is then in **freefall,** which is the state that results when the man and the elevator fall at exactly the same rate. The man's feet will not press on the floor. The scale will read 0 pounds, and he will be floating. You may have noticed this same floating sensation on amusement park rides.

The same thing happens on the Space Shuttle. The Shuttle is constantly falling toward Earth. It just happens to be moving in a circle at the same time. So, just like in the falling elevator, the astronauts in the Shuttle can float because they are falling toward Earth at the same rate as everything else.

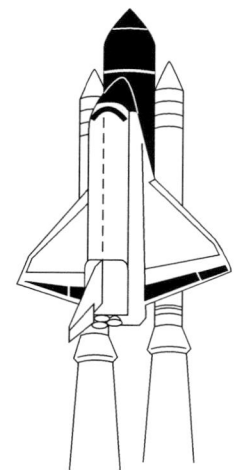

Practice 3—Orbits of Planets and Satellites

Directions: Decide whether each statement that follows is true (**T**) or false (**F**). Write the correct letter in each blank.

_____ 1. The moon travels in a circle around Earth because there are no forces acting on it.

_____ 2. A satellite orbiting Earth is constantly falling, but its rate of falling matches the curvature of Earth.

_____ 3. The speed at which a satellite orbits depends on its distance above Earth.

_____ 4. Satellites must orbit above Earth's atmosphere so that there is no air resistance.

_____ 5. The oval path of orbiting planets is called an ellipse.

_____ 6. Geosynchronous satellites are very close to Earth's surface.

_____ 7. A person in a freely falling elevator would feel "weightless," similar to an astronaut in orbit.

Escape Velocity and Black Holes

The faster you throw a ball upward, the higher up it will go. Would it be possible to throw a ball directly upward fast enough so that it would never come back? It turns out that the answer is yes.

Remember, the farther away you move from Earth, the weaker the gravitational pull. If you throw the ball fast enough, it will get far enough from Earth that gravity won't be able to slow it down enough in order to pull it back. However, you would need to throw the ball at 25,000 miles per hour! This is known as the escape velocity. **Escape velocity** is the velocity that will allow an object to completely escape the gravitational pull of another object.

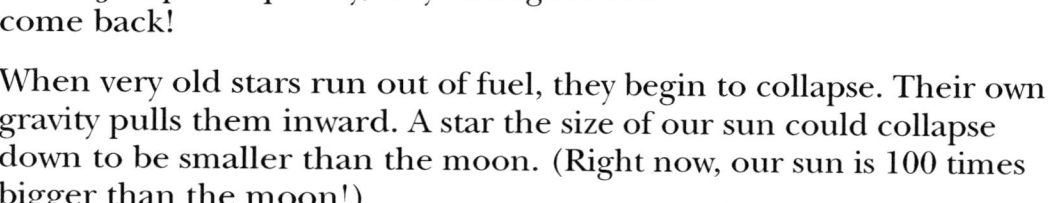

On the moon, the force of gravity is less because the moon has less mass. Therefore, the escape velocity of the moon is only 5,000 miles per hour. On an even smaller object, such as an asteroid, the escape velocity would be even lower. You would have to be careful not to jump too quickly, or you might never come back!

When very old stars run out of fuel, they begin to collapse. Their own gravity pulls them inward. A star the size of our sun could collapse down to be smaller than the moon. (Right now, our sun is 100 times bigger than the moon!)

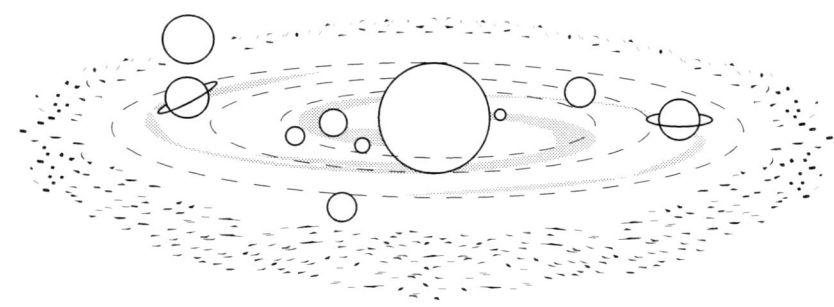

However, for *very* large stars—ones larger than our sun—the mass of the star has such a strong gravitational pull that nothing can stop its collapse. All the mass of the star gets crushed into a tiny point. This tiny point is called a **black hole.**

Because there is so much mass in such a tiny space, the gravitational pull near a black hole is enormous. In fact, the escape velocity near it becomes faster than the speed of light, and no object can *ever* escape!

In the early 1900's, the German scientist Albert Einstein showed that nothing can travel faster than light. This means that no object, even light, has enough speed to escape the black hole. Once an object gets too close, it can never get away. Because not even light can escape, a black hole appears black.

Objects far away from the black hole are not in danger of being pulled in. Just like other objects, the gravitational pull of the black hole decreases as you get farther away. However, at a certain distance away from the black hole, there is a point, known as the **event horizon,** beyond which nothing can ever escape.

If we cannot see black holes, how do we know they exist? The effect of black holes on other objects *can* be seen. Astronomers can observe the effects when stars and other matter are pulled toward a black hole.

Practice 4—Escape Velocity and Black Holes

Directions: Circle the answer that correctly completes each of the following statements.

1. The speed at which an object must be moving in order to completely escape a gravitational pull is called the ____.

    (a) event horizon

    (b) escape velocity

    (c) angular velocity

2. The escape velocity on the moon is ____ on Earth.

    (a) greater than

    (b) less than

    (c) the same as

3. Very close to a black hole, the escape velocity is ____ the speed of light.

    (a) greater than

    (b) less than

    (c) equal to

Directions: Decide whether each statement that follows is true (**T**) or false (**F**). Write the correct letter in each blank.

____ 4. No matter how fast an object is moving, it cannot escape the gravitational pull of Earth.

____ 5. The more mass a planet has, the greater its escape velocity will be.

____ 6. Our sun will one day form a black hole.

____ 7. No matter how far away an object is from a black hole, it cannot escape the gravitational pull of the black hole.

____ 8. Black holes appear black because not even light can escape their gravitational pull.

____ 9. We cannot directly see black holes, but we can see their effect on nearby objects.

■ LESSON MASTERY TEST

Directions: Look at the list of terms below. Write the letter of the correct term in each blank.

(a) centripetal force

(d) acceleration

(b) Galileo Galilei

(e) air resistance

(c) velocity

____ [1] studied the motion of objects falling on Earth. He found that when there is almost no ____ [2], all objects fall with the same constant ____ [3] . As objects fall, their ____ [4] increases by 10 meters / sec each second. Isaac Newton later realized that the same force that causes objects to fall, gravity, provides the ____ [5] that makes the moon travel around Earth.

Directions: Circle the answer that correctly completes each of the following statements.

6. The speed at which an object must be moving in order to completely escape a gravitational pull is called the ____.

    (a) escape velocity

    (b) event horizon

    (c) angular velocity

7. The force of gravity between two objects ____ as they move farther apart.

    (a) increases

    (b) decreases

    (c) remains the same

8. A satellite that always stays directly above one point on the earth is called a ____ satellite.

    (a) translational

    (b) geosynchronous

    (c) ellipse

9. A 60 kg person would weigh ____ on the moon.

    (a) more

    (b) the same

    (c) less

(continued on next page)

10. A golf ball dropped on Earth will fall _____ in 4 seconds. (Remember the formula $d = 1/2\, a \times t^2$. The acceleration of gravity on earth is about 10 m / sec / sec.)

   (a) 10 meters

   (b) 40 meters

   (c) 80 meters

Directions: Decide whether each statement that follows is true (**T**) or false (**F**). Write the correct letter in each blank.

_____ 11. No matter how fast an object is moving, it cannot escape the gravitational pull of Earth.

_____ 12. A dictionary with a mass of 2 kg feels a stronger force of gravity than a text book with a mass of 1 kg.

_____ 13. Astronauts can float in the Space Shuttle because they are in freefall, falling at the same rate as the spaceship they are in.

_____ 14. All satellites orbit Earth at the same speed.

_____ 15. All orbiting planets and satellites move in perfectly round circles.

PART 2

ENERGY AND HEAT

Table of Contents

■ Lesson 1—Mechanical Energy

Goals: To understand work and mechanical energy; to apply the principle of the conservation of energy

Work and Energy

Energy is one of the most important principles in science. It is also one of the most difficult to define or measure. Every object has energy. Any motion or change requires energy. But, we cannot see energy itself.

What we can observe is the effect of energy. In physics, the effect of energy is known as **work.** The scientific meaning of work is somewhat different from the everyday meaning of work. For example, people often say that studying is hard work, or holding up a heavy box is hard work. However, physicists use the term work to describe how much *effect* a force has in causing an object to *move*.

For instance, in the diagram below, the man is only doing work as he lifts up the suitcase. Here, his force is causing the object to move. When he is holding the suitcase still, he is doing no work.

Similarly, if you push against a wall and it doesn't budge, you did not do any work. This is because the force was not effective in causing the wall to move.

The amount of work achieved by a force depends on two things: the amount of force applied and the distance the object moves. For example, in the diagram on page 3, the man would do more work if he lifted the suitcase up over his head because the force would cause the object to move a greater distance. Likewise, if the suitcase was filled with bricks, more work would be done in lifting it because a greater force must be applied.

You can calculate the amount of work done by multiplying the force and the distance, as in the following formula:

$$W = F \times d$$

Here, W is work, F is force, and d is the distance the object moves.

Now, suppose the suitcase weighs 200 N, and the man lifts it up 1.5 meters. The force required to lift the suitcase is 200 N, since the man must overcome gravity to lift it.

$$W = F \times d$$

$$W = 200 \text{ N} \times 1.5 \text{ m}$$

$$W = 300 \text{ N-m}$$

Tip!

Remember: weight is a force, and the metric system uses N to measure force. One pound is about 20 N.

In the metric system, when you multiply Newtons and meters, the unit is known as **joules** (J). Joules are used to measure work. So, the work done is 300 joules.

If the man lifted the suitcase 2.5 meters, then the work done would be:

$$W = F \times d$$

$$W = 200 \text{ N} \times 2.5 \text{ m}$$

$$W = 500 \text{ J}$$

Now, the work done is 500 J.

Although we cannot see energy, **energy** is the ability to do work. The more energy an object has, the more work it can do. In other words, more energy generates a greater force. Or, more energy applies a given force over a greater distance.

Energy and work are both measured in the same units, joules. So, if an object has 300 J of energy, it can do 300 J of work, as in the first example above.

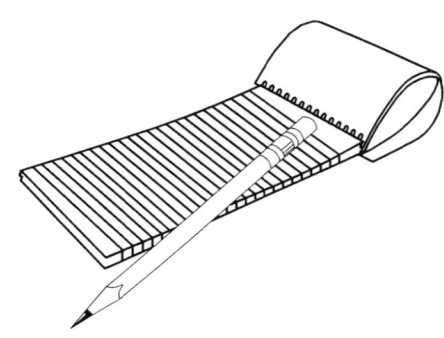

An object can have many kinds of energy: heat energy, electrical energy, chemical energy, mechanical energy, and so on. The same unit of measurement can be used for all of these types of energy.

Tip!

In many calculations, you will use distances, masses, forces, and energy together. In the metric system, the measurements will all be in meters, kilograms, and seconds. Your answer for force will be in Newtons, and the answer for energy will be in joules.

Practice 1—Work and Energy

Directions: Circle the answer that correctly completes each of the following statements.

1. When ____, NO work is done by the force.

    (a) a baseball player hits a baseball

    (b) a person does a push-up

    (c) a weight-lifter holds a barbell stationary over his head

    (d) a man opens a car door

2. Lifting an object 2 meters requires ____ work as lifting it 1 meter.

    (a) twice as much (b) half as much (c) the same work

3. More energy allows an object to do ____ work.

    (a) more   (b) less   (c) the same amount of

4. Energy is measured in units called____.

    (a) Newtons   (b) joules   (c) meters

5. It requires 200 N of force to push a heavy box. You push it for 3 meters. How much work did you do?

    (a) 200 J   (b) 300 J   (c) 600 J

Kinetic and Potential Energy

There are many forms of energy. The easiest form of energy to measure is kinetic energy. **Kinetic energy** is the energy of an object that is moving (also called energy of motion). A falling rock, a speeding bullet, and a child going down a slide are all examples of kinetic energy.

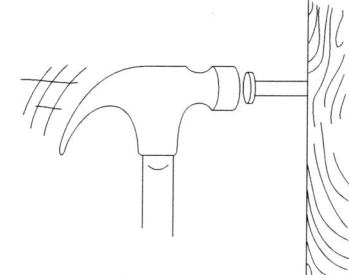

The faster an object is moving, the more kinetic energy it has. Remember that energy is the ability to work. A hammer that is moving quickly can do more work pushing a nail. So, the hammer has more energy. Likewise, the more mass an object has, the more kinetic energy it has. Remember, mass is the amount of matter, or "stuff," in an object. A heavy sledge-hammer can do more work on the nail than a small hammer can.

You can calculate the amount of kinetic energy an object has by using the following formula:

$$K = 1/2 \, m \times v^2$$

Here K is kinetic energy, m is the mass of the object, and v is its velocity.

For example, how much energy is in a 10 kg bowling ball that is rolling at 8 m/sec? (Note that each number and variable that is part of the v ends up being "squared," or multiplied by itself.)

$$K = 1/2 \, m \times v^2$$

$$K = 1/2 \, (10 \text{ kg}) \times (8\text{m/sec})^2$$

$$K = 5 \text{ kg} \times 64 \text{ m}^2 / \text{sec}^2$$

$$K = 320 \text{ N-M} \quad (\text{Remember, N} = \text{kg} \times \text{m/sec}^2.)$$

$$K = 320 \text{ J}$$

So, the bowling ball has 320 J of energy.

An object can also have stored energy, as a result of its position or surroundings. Stored energy is known as **potential energy.** Potential energy is sometimes difficult to measure. Since you can't *see* potential energy in the same way you can see kinetic energy in a moving object, it is harder to identify.

For example, when you compress a spring, stretch a rubber band, or pull back on a bow and arrow, you are storing energy. This energy is released when you let go of the rubber band or bow string. The potential energy of something that is stretched or compressed is known as **elastic** potential energy.

(a)

(b)

The spring (a) can store energy (elastic potential energy) when compressed (b).

Similarly, you store energy in an object when you lift it. For example, you store energy when you lift a hammer. The energy is converted into work when the hammer falls and hits the nail. The energy stored in an object when you lift it up against gravity is known as **gravitational** potential energy.

The amount of potential energy stored in an object can, in fact, be measured. But, you must know how much work was done to store the energy. For example, you must do work to pull back on a bow and arrow. The more work you do, the more energy is stored.

Think About It

How can you increase the work done when pulling back on a bow and arrow? (Remember that work is the amount of force and the distance over which the force is applied.)

Similarly, for gravitational potential energy, you do more work if you lift the object higher. You also do more work if the object's weight is greater. So, the higher the object is, and the greater its weight, the more potential energy it will have.

For example, imagine you drop a baseball on one foot and a bowling ball on the other. The baseball will not hurt, but the bowling ball will. This is because the bowling ball has more stored potential energy. However, if the baseball is dropped from a three-story building, it will hurt when it hits your foot. This is because it has much more potential energy when it is dropped from a greater height. Furthermore, if the bowling ball is dropped on the moon, it will not hurt much at all. On the moon, the weight of the bowling ball is less, so you do less work when you lift it. This means it does not have as much potential energy.

You can calculate the amount of potential energy in an object by calculating the amount of work done to store the energy.

For example, it requires an average force of 50 N to pull back a bowstring. You pull it back 0.25 m (about 10 inches). How much potential energy does the bowstring have?

$$W = F \times d$$

$$W = 50 \text{ N} \times 0.25 \text{ m}$$

$$W = 12.5 \text{ J}$$

You did 12.5 J of work to pull back the bowstring. So, it has 12.5 J of elastic potential energy.

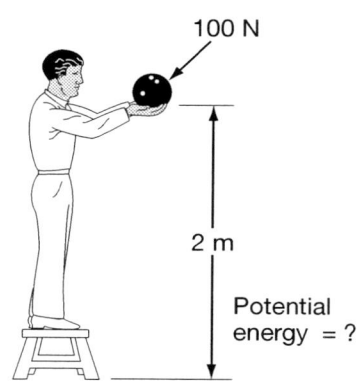

For gravitational potential energy, the force required to lift an object is simply the object's weight. For example, how much gravitational potential energy is in a 100 N bowling ball (about 20 pounds) that is 2 meters above the ground?

$$W = F \times d$$

$$W = 100 \text{ N} \times 2 \text{ m}$$

$$W = 200 \text{ J}$$

You must do 200 J of work to lift the bowling ball. So, it has 200 J of gravitational potential energy.

Practice 2—Kinetic and Potential Energy

Directions: Decide whether each sentence below describes kinetic energy (**K**), gravitational potential energy (**G**) or elastic potential energy (**E**). Write the correct letter in each blank.

_____ 1. A car is driving at 30 mph.

_____ 2. A book is resting on the top of a bookshelf.

_____ 3. A ball rolls across a pool table.

_____ 4. A man stretches his suspenders to put them on.

Directions: Decide whether each statement that follows is true (**T**) or false (**F**). Write the correct letter in each blank.

_____ 5. A golf ball and a bowling ball rolling at the same speed have the same amount of kinetic energy.

_____ 6. The amount of potential energy in an object is equal to the amount of work done to store the energy.

_____ 7. There is only one type of potential energy.

_____ 8. Once energy is stored in an object, the energy cannot be released.

Directions: Circle the correct answer to each of the following questions.

9. If a boy has a mass of 40 kg, and he is running at 5 m/sec, how much kinetic energy does he have? (_Hint:_ Remember the equation $K = 1/2 \, m \times v^2$.)

    (a) 400 J   (b) 500 J   (c) 5,000 J

10. In stretching a rubber band, a person applies 20 N of force over a distance of 0.1 meters. How much potential energy is stored in the rubber band?

    (a) 2 J   (b) 10 J   (c) 20 J

Conservation of Energy

Many of the important ideas in physics are principles known as **conservation laws.** A conservation law says that a quantity of something never changes. Suppose you take a piece of wood and you cut it into tiny pieces. The amount of matter doesn't change. It simply changes its shape. Now suppose you burn it. The matter has changed its form. Now instead of wood, there are ashes, smoke, and gases. However, if you gathered all these together and measured the mass, you would find the same amount of matter. Matter wasn't created or destroyed. The amount of matter didn't change. Only its form changed. We call this principle the **conservation of matter.**

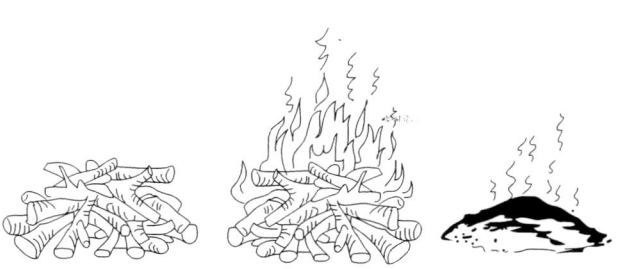

Even though the form of the wood changes as it burns, the amount of matter remains the same.

Similarly, energy cannot be created or destroyed. But, it can change between different forms. The transformation of energy is involved in almost all physical processes, from falling balls to chemical reactions. In all of these processes, the amount of energy after the action is equal to the amount of energy before the action. Energy can change forms. But, the total amount of energy does not change. This principle is known as **conservation of energy.**

We can learn a lot about the transformation of energy by observing different processes. We can also predict the results of these processes. For example, consider a child on a swing. At her highest point, the child has lots of potential energy. This great amount of potential energy is due to her height above the ground. However, she has no kinetic energy because she is momentarily still. As she goes down, the potential energy decreases. This is because the height decreases. However, this energy does not disappear. It changes into kinetic energy as the child speeds up. As the girl on the swing begins to rise, all of her potential energy has changed into kinetic energy. At this point, the speed is greatest because all the energy is in the form of kinetic energy. As she goes back up, the kinetic energy changes into potential energy. She slows down as she rises. When the swing reaches the end of its path, its energy has changed to potential energy again.

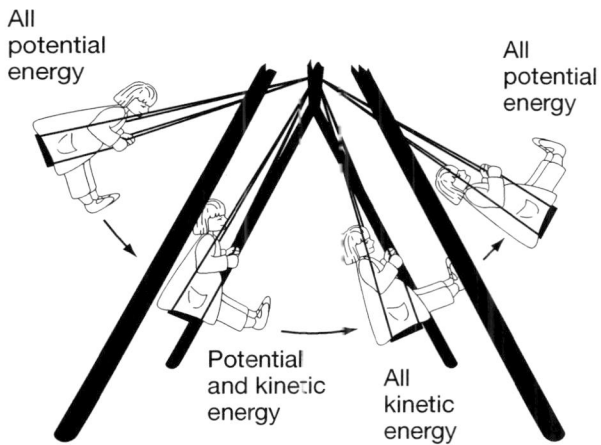

All
potential
energy

All
potential
energy

Potential
and kinetic
energy

All
kinetic
energy

Furthermore, the higher the child swings, the more potential energy she has. Common sense tells you that she will then go faster at the bottom of the swing's path. But, the principle of conservation of energy lets you calculate *how much faster* she will go. You will learn how to do these calculations in the next section.

In some processes, it may appear that energy is lost. For example, when you drop a rock, it has potential energy at first. Then, it has kinetic energy as it is falling. But, when the rock hits the ground, it stops. Its energy appears to be gone. However, remember our example of the burning wood. If you measure only the mass of the ashes, it appears that matter has been destroyed. However, if you consider the mass of the smoke and gases produced, you will find the same amount of matter.

The same idea of conservation holds true for the energy stored in the dropped rock. When the rock hits the ground, it creates a sound. Sound is energy. Likewise, the rock makes the ground vibrate. This is energy also. And finally, the rock hitting the ground creates some heat. If you could measure each of these, and add them up, you would find that the total energy is equal to the amount of energy the rock had before it fell.

In many processes, some or all of the energy is converted into other forms of energy. For example, when a ball is dropped, not all of its potential energy turns into kinetic energy. Air resistance and friction cause some of the energy to be converted into heat. Similarly, without being pushed, a child will not continue to swing forever. Air resistance will cause her energy to slowly decrease, as the mechanical energy is converted into heat. However, when she comes to a stop, the amount of heat created by air resistance will be equal to the energy she had before she started to slow down.

Sometimes, studying the amount of energy that is lost is also interesting. Consider a bicycle rider who coasts down one hill and up another. If no energy was taken away by friction, the bicycle rider would coast up the second hill to the same height at which he began to coast down the first hill. This is because the amount of energy an object has after a process must be equal to the gravitational potential energy it started with. However, some of the bicycle rider's energy is turned into heat because of air resistance and friction in the bicycle wheels. Because of this, the bike has less energy after it descends the first hill, and doesn't coast up to the same height it started from. If the biker makes himself and his bike more aerodynamic (reduces air resistance), less energy will be lost. Then he will coast farther. However, if he applies his brakes, he converts more of his energy into heat. Thus, he will coast less.

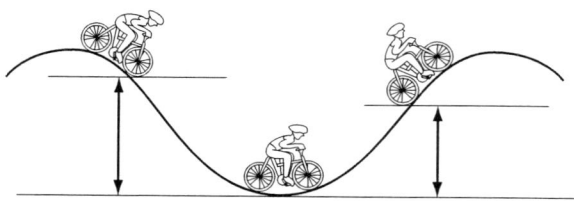

In many sports, conservation of energy is important. A ski jumper converts his potential energy at the top of the hill into the kinetic energy of the jump. Obviously, he can jump farther if he starts higher up the hill. This will give him more energy for the jump. However, he can also jump farther by reducing air resistance. He does this by crouching over to make himself more aerodynamic. He can also wear special clothing which reduces air resistance. Both of these factors will give him more kinetic energy when he takes off. Reducing the friction between the skis and the snow will also reduce the amount of energy loss.

Practice 3—Conservation of Energy

Directions: Decide whether each statement that follows is true (**T**) or false (**F**). Write the correct letter in each blank.

_____ 1. You can destroy matter by burning it.

_____ 2. Although energy can change forms, the total amount of energy does not change.

_____ 3. When an archer fires an arrow, potential energy is converted to kinetic energy.

_____ 4. When a bicyclist puts on the brakes, energy is destroyed.

_____ 5. Friction removes energy from an object, but does not destroy it. The energy is transformed into heat.

_____ 6. The law of conservation of energy can be used to predict the results of physical actions, since the total energy after the action must be equal to the total energy before.

Calculating with Conservation of Energy

Conservation of energy applies to everything in nature. Therefore, it is a powerful tool to help you make predictions. If you know the energy before something happens, the energy afterwards must be equal. For example, when you pull back on a bow and arrow, you store energy. When they are released, the energy changes forms, but the total amount doesn't change. If you know the energy of the arrow, you can predict how fast and how far it will go. The same principles can be used to predict the result of a chemical reaction, the speed of a spacecraft in outer space, or how high a child will go on a swing.

For example, a juggler throws a ball that weighs 60 N directly upwards. He throws it fast enough so that it has 300 J of energy. How high will it go?

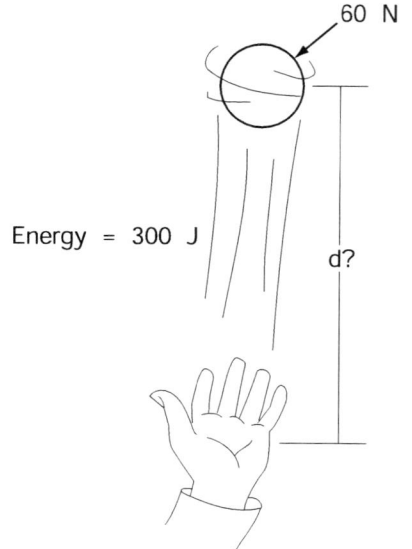

As the ball rises, its kinetic energy will turn into potential energy. At the top of its path, all 300 J will have turned into potential energy. Now, you just need to calculate the height that 300 J of potential energy represents.

$$PE \text{ (potential energy)} = F \times d$$

Potential energy is the force (F) required to move the object times the distance (d) it moved (in this case, the height).

$$300 \text{ J} = 60 \text{ N} \times d$$

Now, divide both sides of the equation by 60 N to find the distance the ball goes up.

$$\frac{300 \text{ J}}{60 \text{ N}} = \frac{60 \text{ N} \times d}{60 \text{ N}}$$

$$d = 5 \text{ m}$$

(Remember that J is the same as N-m.)

So, the ball will rise 5 meters.

Let's look at another example. An archer fires an arrow that has a mass of 0.1 kg into a target. The arrow is moving at 30 m/sec.

(a) How much kinetic energy does the arrow have?

KE (kinetic energy) $= 1/2\ m \times v^2$

$KE = 1/2\ (0.1\ \text{kg}) \times (30\ \text{m/sec})^2$

$KE = 1/2\ (0.1\ \text{kg}) \times (900\ \text{m}^2/\text{sec}^2) = .05\ \text{kg}\ (900\ \text{m}^2/\text{sec}^2)$

$KE = 45\ \text{J}$

(b) How much work can the arrow do when it hits the target?

The arrow can use all 45 J of kinetic energy to push its way into the target.

(c) If it requires 225 N to push through the target, how far into the target will the arrow go?

The arrow can do 45 J of work, so you can use the equation for work to determine what distance it moves.

$W = F \times d$

$45\ \text{J} = 225\ \text{N} \times d$

Divide both sides of the equation by 225 N to find the distance the arrow will go into the target.

$$\frac{45\ \text{J}}{225\ \text{N}} = \frac{225\ \text{N} \times d}{225\ \text{N}}$$

$d = 0.2\ \text{m}$

It will go 0.2 m into the target.

Tip!

All conservation of energy problems require you to find the energy beforehand (if it is not given to you in the problem). The total energy afterward will be the same. First, figure out what form the energy takes. Then, use the appropriate equation to solve for whatever value is asked for.

Of course, the problems that physicists and engineers solve using conservation of energy are often much more complicated than these. However, they utilize the same idea. The total energy never changes, so if you can keep track of where the energy goes, you can predict the result of the process.

Practice 4—Calculating with Conservation of Energy

Directions: Read the following problems. Using the formulas you just learned, write the correct answer in each blank.

A wagon with a mass of 20 kg is moving with a speed of 5 m/s.

_____ 1. How much kinetic energy does it have?

_____ 2. How much work was done to give it this energy?

_____ 3. If the wagon was pushed for 10 meters, how much force did it need to be pushed with to give it this energy? (*Hint:* Use the equation for work: $W = F \times d$.)

A boy is jumping on a pogo stick. His weight is 500 N, and he jumps up 0.3 meters.

_____ 4. How much gravitational potential energy does he have at the top of his jump? (*Hint:* $PE = F \times d$.)

_____ 5. How much work can he do to compress the pogo stick's spring when he lands?

_____ 6. If it takes an average force of 750 N to compress the spring, how far will it compress? (*Hint:* This is similar to question 3 above.)

■ LESSON MASTERY TEST

Directions: Decide whether each sentence below describes kinetic energy (**K**), gravitational potential energy (**G**), or elastic potential energy (**E**). Write the correct letter in each blank.

_____ 1. A car drives at 30 miles per hour.

_____ 2. A person sets a spring-loaded mousetrap.

_____ 3. A bungee jumper stands at the top of a bridge.

_____ 4. A bungee jumper falls through the air at 30 m/sec.

_____ 5. A bungee jumper stretches the cord by 10 meters at the bottom of her fall.

Directions: Circle the correct answer to each of the following questions.

6. A 3 kg rock is falling at 10 m/sec. How much kinetic energy does it have?

   (a) 30 J

   (b) 45 J

   (c) 150 J

7. It requires 150 N average force to pull a bowstring back 0.2 meters. How much energy is stored in the bowstring?

   (a) 20 J

   (b) 30 J

   (c) 40 J

8. In which of the following cases is NO work done by the force?

   (a) A man lifts a heavy suitcase into a car.

   (b) A woman holds her child in her arms.

   (c) A car accelerates to 60 mph in 10 seconds.

   (d) A boy pushes his bicycle through some mud.

(continued on next page)

Directions: Decide whether each statement that follows is true (**T**) or false (**F**). Write the correct letter in each blank.

_____ 9. You can destroy matter by cutting it into small pieces.

_____ 10. A falling rock gains energy as it moves faster.

_____ 11. Energy is the capacity to do work.

_____ 12. It takes half as much work to lift a box 2 meters as it does to lift the box 1 meter.

Directions: Read the following problem. Using the formulas you have learned, write the correct answer in each blank.

A 20 kg ball is thrown into the air with 140 joules of kinetic energy.

_____ 13. At the top of its path, how much potential energy will it have?

_____ 14. How high will it be at the top of its path? (_Hint:_ Remember that $PE = F \times d$.)

■ Lesson 2—Temperature and Heat

Goals: To understand the physical meaning of temperature and heat; to understand the effect of temperature and heat on objects

Temperature

Your sense of touch can easily tell you if an object is hot or cold. Using a thermometer, you can measure the temperature of the air around you. However, it wasn't until the 19th century that scientists really understood what temperature is.

All matter is made up of atoms and molecules. However, these atoms and molecules are not stationary. Even in a solid, they are continually jiggling and moving in place. In a liquid or gas, the molecules are even more free to move around. When you heat up a substance, the molecules begin to move faster. **Temperature** is perceived as the increased motion of molecules in a substance. In a glass of cold water, the water molecules are moving more slowly than in a glass of hot water. On a warm summer day, the air molecules are moving faster than on a cold winter day.

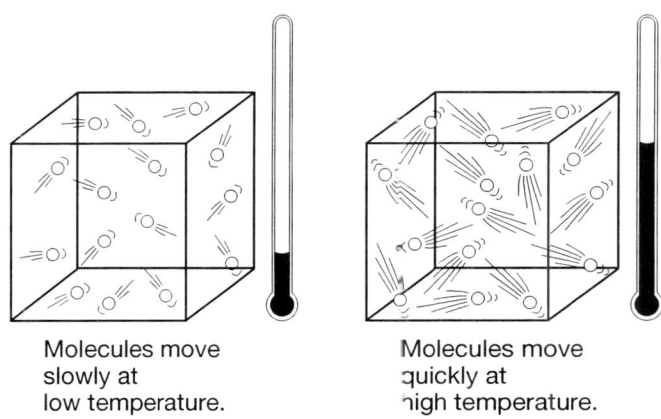

Molecules move slowly at low temperature.

Molecules move quickly at high temperature.

Remember that kinetic energy is the energy of motion. Therefore, the molecules have kinetic energy. The faster they move, the more kinetic energy they have. **The temperature of an object corresponds to the average kinetic energy of its molecules.**

The two most common scales used to measure temperature are the **Fahrenheit** and **Celsius** scales. Both of these are based on the freezing point and boiling point of water. On the Fahrenheit (F) scale, water freezes at 32°F and boils at 212°F. The Celsius (C) scale makes the freezing point of water 0°C, and the boiling point of water 100°C. The Fahrenheit scale is commonly used in the United States. The Celsius scale is used in most of the rest of the world. Most scientists also use the Celsius scale.

More recently, after scientists better understood the molecular basis of temperature, another scale was invented. This scale is known as the **Kelvin** (K) scale. On this scale, 0° corresponds with zero kinetic energy. This makes sense because temperature represents the kinetic energy of molecules. But, is it possible to have zero energy?

As you cool a substance, its molecules move more and more slowly. If you could remove all the energy, the molecules would nearly stop moving. This is the lowest possible temperature, known as **absolute zero.** It occurs at −273°C or −459°F.

The Kelvin scale locates absolute zero at 0 K. Each degree on the Kelvin scale is the same as one degree on the Celsius scale. On the Kelvin scale, water freezes at 273 K and boils at 373 K. The Kelvin scale isn't used much outside of science. A cool day is around 270 K, while a hot day measures around 300 K. These high numbers make the Kelvin scale cumbersome for everyday use. However, this scale is very useful in physics, since zero degrees corresponds with zero energy.

Absolute zero	Water freezes	Room temperature	Water boils
0 K	273 K	293 K	373 K
−273°C	0°C	20°C	100°C
−459°F	32°F	70°F	212°F

Almost all forms of matter expand when their temperatures increase. Matter also contracts when its temperature decreases. A metal bar becomes longer when heated. A balloon full of air will expand slightly on a hot day. This is known as **thermal expansion.**

Fill a balloon with air. Place it in your freezer. When you take it out an hour later, it will be smaller. This is because the air inside the balloon cools down and contracts.

When the temperature of a substance increases, the molecules in that substance start jiggling faster. This pushes the molecules apart from each other, making the substance expand.

To better understand the concept of thermal expansion, imagine a line of children standing next to each other. If they stand still, the line will be short. If they start bumping into each other, they push apart. This makes the line become longer. The same thing happens when a metal bar is heated.

Thermal expansion has many important applications. For example, a thermometer works on the principle of thermal expansion. Inside the glass tube of a thermometer is a liquid, usually alcohol or mercury. When the thermometer is hot, the liquid expands. This makes it rise up inside the tube, indicating a higher temperature. The greater the temperature, the more the liquid expands.

You may have seen a glass crack when a hot drink was poured in it. This is because when the drink hits the bottom of the glass, the temperature of the bottom increases. The bottom of the glass expands before the top does. This uneven expansion causes the glass to crack. High quality cookware is designed either to expand very little when heated, or to be strong enough to withstand the uneven expansion.

Often, engineers have to take thermal expansion into account when designing bridges or buildings. A long bridge expands in the summer, so small spaces are left between segments of the bridge. As the bridge expands, it fills in these gaps.

Water is an unusual substance. When it freezes, it expands rather than contracts. If you've ever put a can or bottle of liquid in the freezer, you may have found it broken later. As the water inside freezes, it expands and takes up more space, breaking the container. This is why cars have anti-freeze in the coolant. The anti-freeze prevents the coolant from freezing, because if it did, it would expand and crack the engine.

Practice 1—Temperature

Directions: Circle the answer that correctly completes each of the following statements.

1. Temperature is a measure of the average ____ of the molecules in an object.

    (a) kinetic energy

    (b) potential energy

    (c) mass

2. At higher temperatures, the molecules in a substance move ____.

    (a) faster

    (b) slower

    (c) in a curved path

3. The ____ scale sets zero degrees equal to zero kinetic energy.

    (a) Celsius   (b) Fahrenheit   (c) Kelvin

4. ____ is the lowest possible temperature, where molecules have no random motion.

    (a) Absolute zero

    (b) Thermal zero

    (c) Absolute power

5. When ice freezes, it ____.

    (a) expands

    (b) contracts

    (c) increases its kinetic energy

6. A metal bridge ____ during summer.

    (a) contracts   (b) expands   (c) doesn't change

7. If you ever have difficulty unscrewing the metal lid of a jar, try running hot water over the lid. This makes the lid easier to open because ____.

    (a) the jar contracts

    (b) the lid contracts

    (c) the lid expands

Heat

Heat is energy that is transferred because of a difference in temperature between two objects. For example, when you put your hand on a piece of hot metal, your hand feels hot. This is the energy moving into your hand, making it hot. When you put food in the refrigerator, heat leaves the food and goes into the air of the refrigerator. This is how the food cools off.

Heat always moves from hot objects to cold objects. For example, if your hands are cold, and you put them in warm water, you will never find that your hands remain cold and the water becomes cooler.

When a hot object touches a cold object, the molecules in the hot object are moving much more quickly. The molecules in the hot object bump into the molecules in the cold object. They transfer some of their energy. This makes the molecules in the cold object move faster. If the two objects are in contact long enough, they will eventually reach the point where the molecules in each have the same kinetic energy. Therefore, the objects will have the same temperature. At this point, no more heat will be transferred. The two objects are now in **thermal equilibrium.**

For example, consider a glass of iced tea left outside on a hot day. The quickly moving hot air molecules will collide with the cold molecules in the glass. This action transfers heat into the glass. The glass of iced tea will continue to warm up until it reaches thermal equilibrium. That is when it becomes the same temperature as the air around it.

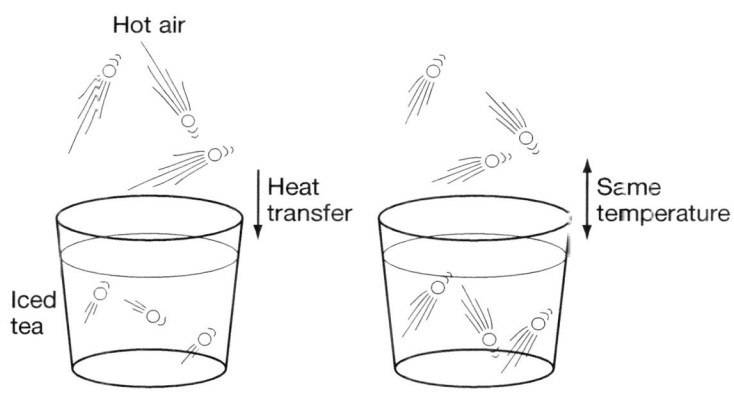

Thermal Equilibrium

Heat, like other forms of energy, can be measured in joules. However, heat is also often measured in calories. One **calorie** is the amount of heat required to raise the temperature of 1 gram of water by 1°C. One calorie is equal to about 4 joules. The **Calories** (capital C) used to measure food energy are actually **kilocalories.** One food Calorie is equal to 1,000 calories.

When you add heat to a substance, its temperature increases. However, the amount that the temperature changes depends on two factors. The first factor is the quantity of the substance that is being heated. It takes a lot more heat to increase the temperature of a large pot of water than it does to increase the temperature of a small pot of water.

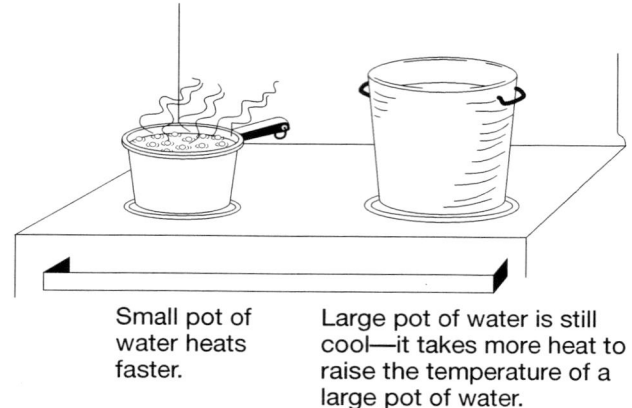

Small pot of water heats faster.

Large pot of water is still cool—it takes more heat to raise the temperature of a large pot of water.

The second factor is the type of substance being heated. Even if you have the same amount of two substances, their change in temperature will be different. For example, 1 calorie will raise 1 gram of water by 1°C. But, this same calorie will raise the temperature of 1 gram of aluminum by 5°C.

To describe this difference, scientists use specific heat. The **specific heat** of a substance is the amount of heat needed to change the temperature of 1 gram of the substance by 1°C. For example, the specific heat of water is 1 calorie, while the specific heat of aluminum is ⅕ calorie. This means it takes only ⅕ the amount of heat to change the temperature of aluminum as it does to change the temperature of water.

Water actually has one of the highest specific heats among common substances. This is why the temperature of the ocean doesn't change as much during the day as the temperature of the air. Water has a higher specific heat than air. This is why much more heat is required to change the temperature of water than to change the temperature of air. This also explains why areas near the ocean are warmer during the winter. The temperature of the ocean doesn't change much. Therefore, the ocean temperature moderates the temperature of the air near the coast. Water is also very useful as a coolant. It can absorb a lot of heat without changing its temperature significantly.

You can calculate the amount of heat needed to change the temperature of a substance by using the following formula:

$$Q = m \times c \times T$$

Here Q is heat in calories (physicists use the letter Q for heat, because H is used for another quantity you haven't learned about); m is the mass of the substance; c is the specific heat of the substance; and T is the change in temperature.

Let's look at an example. How much heat is required to raise the temperature of a 200-gram pot of water from 20°C to 100°C?

The temperature changes by 80°C, and the specific heat of water is 1 cal/gm°C, so:

$$Q = m \times c \times T$$

$$Q = 200 \text{ gm} \times 1 \text{ cal/gm°C} \times 80°C$$

$$Q = 16,000 \text{ calories}$$

So, 16,000 calories are needed. Or, since one food Calorie is equal to 1,000 calories, it would take 16 food Calories to raise the temperature of the water by 80°C.

Practice 2—Heat

Directions: Circle the answer that correctly completes each of the following statements.

1. Heat is energy that is transferred due to a difference in ____.

   (a) mass   (b) temperature   (c) force

2. When an object is at thermal equilibrium, its temperature is ____ the temperature around it.

   (a) less than   (b) more than   (c) the same as

3. Heat can be measured in ____.

   (a) joules   (b) calories   (c) joules or calories

4. ____ is needed to raise the temperature of 1 gram of water by 1°C.

   (a) 1 calorie   (b) 1 joule   (c) 1 pound

5. ____ are needed to raise the temperature of a 10 gram aluminum block by 30°C. (Remember $Q = m \times c \times T$. The specific heat of aluminum is 0.2 cal/gm°C.)

   (a) 30 calories   (b) 60 calories   (c) 600 calories

Directions: Decide whether each statement that follows is true (**T**) or false (**F**). Write the correct letter in each blank.

_____ 6. Heat always moves from cold objects to hot objects.

_____ 7. The specific heat of a substance is the amount of heat required to raise the temperature of 1 gram of that substance by 1°C.

_____ 8. Coastal cities are warm because the ocean water heats up quickly.

Phase Changes

There are three common states, or **phases,** of matter that you see every day—solids, liquids, and gases. For example, water can be a liquid, as in drinking water. It can be a solid (ice), and it can be a gas (steam). You can cause a substance to change from one phase to another by adding or removing heat. This is known as a **phase change.**

Suppose you put a piece of ice that has been in the freezer at $-10°C$ into a pan on the stove. As the heat from the stove is transferred into the ice, the ice in the ice start to wiggle around a little more. The temperature of the ice is increasing. However, the molecules will remain tightly bonded to each other until the ice reaches $0°C$. Remember, $0°C$ is the melting point of water. At this point, the energy that is added doesn't make the molecules jiggle more. The additional energy actually starts to break the bonds between the molecules. Now, they can move freely as a liquid. This is a phase change. The solid ice changes to liquid. The temperature of the water will not increase until enough energy has been added to break all the bonds.

This additional heat energy is called the **latent heat of fusion.** This latent heat of fusion does not increase the temperature. It only breaks the bonds between molecules. Latent means "hidden." This energy is not visible as an increase in temperature. It goes into the invisible process of breaking the bonds.

Once all the ice has melted, the temperature will begin to increase again. The loose water molecules start moving faster. The water begins to boil. Once the temperature reaches $100°C$, some of the molecules can break free of the loose attraction that holds them together as a liquid. These molecules form a gas. This gas escapes into the air as steam. Again, the temperature of the boiling water will not increase above $100°C$ until all the molecules have broken free and become a gas. The extra heat needed to turn a liquid into a gas is known as the **latent heat of vaporization.**

The latent heat in phase changes has several important consequences. Extra energy must be supplied to change a solid to a liquid, and a liquid to a gas. This energy is then released in the opposite direction, from gas to liquid, or from liquid to solid. During these transitions, the temperature remains constant.

For example, it is possible to boil water in a paper cup. You might think the paper would catch fire. But, the temperature of the water will not rise above 100°C until it is completely boiled away. The water keeps the paper cup at 100°C also. Paper does not burn until it reaches 233°C.

Water is very effective for quenching a fire. It prevents oxygen from reaching the flame. It also cools the flame. The water hits the fire and turns to steam. It absorbs a large latent heat of vaporization and takes heat away from the fire.

You may have noticed that you feel cold when you get out of a pool and don't dry off immediately. As the water on your skin evaporates, it also absorbs a large latent heat of vaporization. This latent heat of vaporization is from your body. That is what cools you off. The reverse process occurs when you sit in a sauna. The steam in the air condenses on your skin. This releases the latent heat, and warms your skin.

You can test the temperature of a hot surface, such as a clothes iron, by wetting your finger and then touching the surface. A large amount of heat must be absorbed to vaporize the moisture on your finger. Therefore, the heat is absorbed by the moisture, rather than burning your finger. The same principle is used by "fire-walkers." They wet their feet before walking over hot rocks. The heat has to vaporize the water before it can burn their feet.

The boiling and freezing points of water are actually not constant. They both depend on the pressure surrounding a substance. For instance, the freezing point in a block of ice gets lower as you apply pressure to it. This happens when you ice skate. The blade of the ice skate presses on the ice. This causes the ice to melt. The skate then glides on this thin layer of water. When you press snow together to make a snowball, the pressure you apply momentarily melts the snow. When you release the pressure, the snow re-freezes into a more solid ball.

Likewise, the boiling point of water depends on the atmospheric pressure around it. Lower atmospheric pressure makes it easier for the molecules to escape into the air. So, in places with less atmospheric pressure, such as on top of a mountain, water boils at a lower temperature. For example, in Denver, which is located at an altitude of 5,000 feet, water boils at only 95°C. As you have learned, water cannot rise above the boiling point until it has completely boiled. So, the temperature of liquid water can't increase above 95°C. Therefore, it takes longer to cook foods that must be boiled. The opposite effect occurs in a pressure cooker. The increased pressure in the cooker raises the boiling point. This allows water to remain in liquid form at higher temperatures, and makes the food cook faster.

Evaporation is a slow form of vaporization. You may have noticed evaporation if you have left a glass of water out for a long time. In any substance, not all the molecules move at the same speed. Some move quickly. Some move slowly. Temperature measures the *average* speed. So, some water molecules can spontaneously move fast enough to escape the liquid into the air. This causes evaporation.

Only three states of matter are commonly observed: solid, liquid, and gas. But, if a substance is hot enough, it will enter a fourth phase. This phase is known as **plasma.** In a plasma, the substance is so hot that the speed of the atoms causes the atoms to collide hard enough to shake the electrons away from the nucleus. Instead of atoms bouncing around, the electrons and nuclei move around separately. This occurs only at extremely high temperatures, such as those found at the center of the sun.

Neon signs contain neon gas in a plasma state. An electric voltage causes the electrons to be pulled away from the neon atoms, creating a plasma which glows. The Northern Lights, also known as the aurora borealis, are also the result of a plasma in the upper levels of the atmosphere.

Practice 3—Phase Changes

Directions: Decide whether each statement that follows is true (**T**) or false (**F**).
Write the correct letter in each blank.

_____ 1. It is impossible for most substances to change from a solid to a liquid.

_____ 2. Once water is raised to 100°C, it will instantly turn into steam.

_____ 3. Energy is required to break the bonds between the molecules of a solid when it melts.

_____ 4. Heat is released when steam condenses into liquid water.

_____ 5. The boiling temperature of water is the same everywhere.

_____ 6. You can melt ice by applying pressure to it.

_____ 7. If a substance is cold enough, it will turn into a plasma.

_____ 8. In a plasma, the electrons are removed from the nucleus by collisions between atoms.

Transfer of Heat

You have already learned that heat flows naturally from hot objects to cold objects. In this section, you will learn the three ways that heat can move.

Conduction is the simplest form of heat transfer. **Conduction** occurs when two objects, or different parts of one object, are directly touching each other. Then the molecules in the two objects can collide. This directly transfers kinetic energy.

For example, when you hold a metal spoon in a pan of boiling water, the quickly moving water molecules collide with the metal molecules in the spoon. This makes the molecules in the spoon move faster. These molecules then collide with molecules farther up in the spoon. These collisions keep transferring the heat up the spoon. Eventually, the metal molecules at the top of the spoon collide with the molecules in your hand. The heat is finally transferred to your hand.

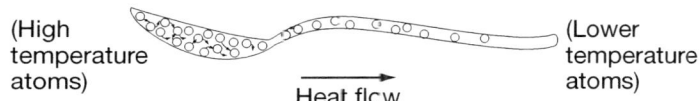

(High temperature atoms)   Heat flow   (Lower temperature atoms)

Heat Transfer by Conduction

Some substances conduct heat better than others. You may have noticed that a plastic spoon does not heat up as much as a metal spoon. A substance, such as metal, that is good at conducting heat is known as a heat **conductor.** Plastic is a poor conductor of heat. That's why the heat does not travel through the plastic spoon into your hand.

You may have also noticed that on a cold day, a metal pole feels very cold, while a wooden pole does not feel cold. This is not because the metal pole is at a lower temperature. In fact, both poles are at the same temperature. However, the metal is a better conductor, so it can absorb the heat from your hand more quickly. This makes it feel colder than the wooden pole feels when you touch it.

Think About It

Malik baked a cake in an oven set at 350°F. When the cake was done, everything in the oven was at 350°F. Malik noticed that when he touched the air inside the oven, it did not feel uncomfortably hot. His cake felt hot to the touch, but it did not seriously burn him. But, Malik knew that if he touched the metal cake pan, he would quickly be burned. Why?

In general, solids are the best conductors. Gases are the poorest conductors. Because the molecules in a solid are close to each other, they can easily collide and transfer heat. On the other hand, the molecules in a gas are far apart. So, they don't collide often, and don't transfer heat as well.

Gases that are the poorest conductors of heat will generally make the best insulators. An **insulator** is a material that prevents the transfer heat energy. Styrofoam makes a good insulator for coffee cups because it contains a lot of trapped air. This trapped air is a good insulator. Similarly, goose down (small feathers) makes good insulation for blankets and jackets. Lots of air can be trapped between the feathers.

Of course, the best insulation occurs when there are *no* molecules to conduct heat. This is used in some types of thermos bottles. Inside the walls of the thermos, the air has been almost completely removed. This leaves a vacuum. With few molecules to conduct the heat, almost no energy can move into or out of the thermos. This allows substances in a thermos to remain cold or hot for long periods of time.

An important concept to understand is that "cold" is not a physical quantity. While heat is energy, which can be transferred, "cold" is simply the absence, or lack, of heat. In terms of physics, "cold" is not transferred between objects. When you touch a piece of ice, it feels cold because heat from your hand enters the ice, not because "cold" from the ice enters your hand. Similarly, using a blanket on a cold night does not keep any substance called "cold" out. It simply keeps in the heat that is produced by your body.

The next form of energy transfer is convection. **Convection** occurs when a moving substance, such as water or air, carries heat along with it. For example, you do not have to touch an electric heater in order to feel its warmth. The heater raises the temperature of the air around it, and the air circulates through the room, carrying the heat with it. Eventually, the heat is spread throughout the room.

Convection often makes use of the fact that hot air rises. Remember that hot air expands. It becomes less dense than the cold air around it. So, it floats upward, as in a hot-air balloon. For example, most heaters are placed on the floor, so that as the hot air rises, it fills the entire room. A heater on the ceiling would not be as effective because the hot air would stay at the top of the room and leave cold air below it.

Likewise, when you boil a pot of water on the stove, the flame directly heats only the water at the bottom of the pot. However, the heated water rises and carries its heat to the top of the pot. The heated water is then released into the unheated water at the top.

Heat Transfer by Convection

A convection oven uses the principle of convection to cook food more quickly. Rather than allowing the heat to slowly conduct through the air to the food, a fan inside the oven causes the air to circulate. This circulation carries the heat directly to the food.

The final form of energy transfer is radiation. **Radiation** uses electromagnetic waves (light, infrared, ultraviolet, etc.) to carry energy. For example, the sun's heat does not reach us by conduction (Earth does not directly touch the sun). We do not feel the sun's heat by convection (hot air from the sun does not travel through space to Earth). Instead, light and other forms of electromagnetic (EM) radiation carry heat with them. You absorb this radiation when you sit out in the sun.

In fact, all objects emit, or put out, some heat by radiation. Objects around room temperature emit mostly infrared radiation. Higher temperature objects, such as a fire or the sun, put out visible light and even higher frequency waves such as ultraviolet.

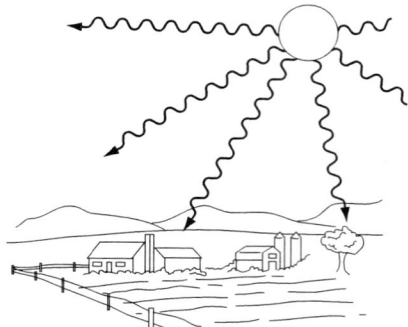

Heat Transfer by Radiation

For example, most of the hot air from a fire in a fireplace rises up the chimney. This takes heat away by convection. The only heat that reaches you directly comes through the radiation put off by the fire.

At night, Earth cools off as it emits heat radiation into space. However, if it is cloudy, some of the radiation is reflected back down to Earth. Therefore, Earth doesn't cool as much on cloudy nights. The coldest nights are usually clear nights.

Different objects absorb heat radiation at different rates. For example, black objects absorb almost all of the radiation that hits them. Light objects reflect almost all of the radiation that hits them. Therefore, dark-colored objects heat up more quickly than light-colored objects when left in the sunlight.

Practice 4—Transfer of Heat

Directions: Decide whether each sentence below describes conduction (**A**), convection (**B**), or radiation (**C**). Write the correct letter in each blank.

_____ 1. You burn your tongue eating a slice of hot pie.

_____ 2. A thermometer reads a higher temperature when it is in direct sunlight.

_____ 3. A fan blows cool air through a room.

_____ 4. Your skin feels warm when you hold it near a light bulb.

_____ 5. You cool off a glass of lemonade by putting ice in it.

_____ 6. A heater on the floor heats your entire house as the hot air rises.

Directions: Decide whether each statement that follows is true (**T**) or false (**F**). Write the correct letter in each blank.

_____ 7. A down blanket keeps you warm by creating heat.

_____ 8. Gases are poor conductors, because the molecules are so far apart.

_____ 9. Dark-colored objects absorb more radiation than white objects.

_____ 10. All objects emit some heat radiation.

■ LESSON MASTERY TEST

Directions: Circle the answer that correctly completes each of the following statements.

1. Temperature is a measure of the average _____ of the molecules in an object.

    (a) kinetic energy   (b) potential energy   (c) mass

2. It requires _____ heat to boil 2 liters of water than to boil 1 liter of water.

    (a) more   (b) less   (c) the same amount of

3. _____ is the extra heat needed to break the bonds when a solid melts.

    (a) Latent heat of vaporization

    (b) Latent heat of fusion

    (c) Thermal equilibrium

4. The temperature of boiling water at sea level is always _____ 100°C.

    (a) less than   (b) more than   (c) equal to

Directions: Decide whether each statement that follows is true (**T**) or false (**F**). Write the correct letter in each blank.

_____ 5. Air is a good conductor of heat.

_____ 6. A liquid's temperature is constant when it is boiling.

_____ 7. Heat always moves from cold objects to hot objects.

_____ 8. It is easier to burn yourself on a hot metal spoon than on a hot wooden spoon, because metal is a better conductor of heat.

_____ 9. Water boils at a lower temperature at high altitudes.

Directions: Decide whether each sentence below describes conduction (**A**), convection (**B**), or radiation (**C**). Write the correct letter in each blank.

_____ 10. Inside some ovens, a fan blows hot air directly onto the food.

_____ 11. Your hand feels cold when you hold a snowball.

_____ 12. An air conditioner blows cold air into your room.

_____ 13. You dry your clothes by placing them in a sunny location.

_____ 14. You fry an egg in a metal pan.

■ Lesson 3—Heat Engines and Thermodynamics

Goals: To understand the functioning of a heat engine; to understand how the laws of thermodynamics and entropy limit a heat engine's performance

Heat Engines

In the first lesson in this book, you studied many processes in which mechanical energy was converted into heat. For example, you learned that when a bicyclist coasts to a stop, his kinetic energy is converted into heat by air resistance and friction. Is it possible to reverse this process, and use heat to make something move?

Heat engines do exactly this. A **heat engine** is a device that uses heat, such as from steam or burning fuel, to create motion. A heat engine converts heat into mechanical work. An internal combustion engine, a steam engine, and a jet engine are all examples of heat engines.

The diagram to the right will help you understand what a heat engine does. The diagram shows two containers filled with gas. The container on the left has immovable walls. The container on the right has a movable piston. When the gas inside the left container is heated, the temperature will increase. This makes the molecules move faster. The molecules collide into the walls with more force. This exerts more pressure on the walls as the gas tries to expand. However, since the walls can't move, the only effect is an increase in the temperature and pressure of the gas. This container will not function as a heat engine.

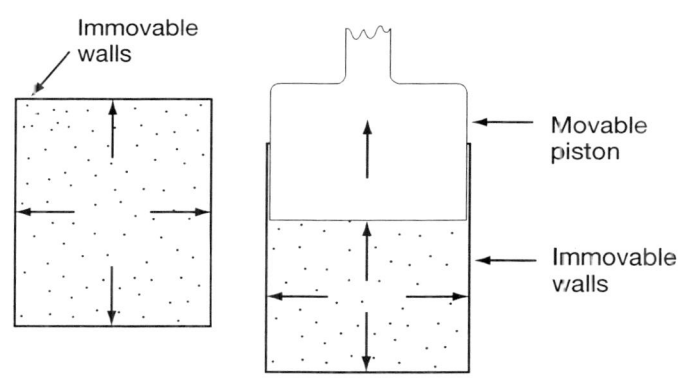

When the gas in the container on the right is heated, the same thing happens. The molecules collide faster and the pressure increases. The pressure of the gas exerts a force which causes the piston to move upward. This container shows what a heat engine does.

Remember that work must be done if a force causes an object to move. So, the gas performs work as it pushes the piston. Where did the energy come from to do this work?

Each time a molecule collides into the piston, it loses some of its kinetic energy. This is just like throwing a tennis ball against a soft surface. The tennis ball loses some of its energy. So, as the molecules push the piston, they slow down. In other words, the temperature decreases. Some of the heat energy is used to perform the work. This makes the temperature decrease as it does work.

Whenever a gas pushes against something to expand, some of the heat energy of the gas must be converted into work, so its temperature decreases. Thus, the temperature decreases as energy is used. This is known as an **adiabatic expansion.**

You can see this effect by blowing on your hand. If you blow with your mouth wide open, the air feels warm. If you almost close your lips and then blow, the air feels cool. When your lips are almost closed, the air must perform work in order to escape and expand. So, it loses some of its energy, and becomes cooler.

Adiabatic expansion also affects weather. As air rises, it expands. As it adiabatically expands, it cools off. This is part of the reason why it is so much colder on top of a mountain than it is in a valley.

On the other hand, think about the container with the piston that we just discussed. If you pushed on that piston with your hand to compress the gas, you would be doing work on the gas. The work done in compressing the gas is stored as heat energy. This increases the temperature of the gas. So, when a gas is forced to compress, its temperature increases. This temperature increase and storing of energy is known as **adiabatic compression.** You may have noticed the heating due to adiabatic compression when using a bicycle pump. The pump feels very warm after you use it. This is because of all the work done to compress the air inside.

So, in the process of adiabatic expansion, some heat energy is used to perform work. However, in order to have a useful heat engine, this process must repeat continually. It doesn't do any good to have an engine which just pushes once!

In a heat engine, the piston has to move back down and repeat the cycle. This can be done in one of two ways. One way is for the gas in the engine to be cooled from the outside so that it contracts. Another way is for the gas to be heated from the inside. This creates pressure, which is lessened when the gas is released from the container as exhaust.

An internal combustion engine, as in an automobile, uses the second technique. The internal combustion, or gasoline engine works by burning (igniting) gasoline in a cylinder containing a piston as shown below. Gasoline and air are mixed and drawn into a cylinder during the **intake stroke.** The gasoline mixture is ignited by a spark during the **ignition stroke.** The heat that is produced causes the air in the cylinder to expand, pushing the piston down. The piston is connected to a crankshaft. When the piston moves down, the crankshaft turns. Then, the burnt gases are released during the **exhaust stage.** This allows the piston to move back up, again turning the crankshaft.

The turning crankshaft is how the back-and-forth motion of the piston is converted to rotate a wheel.

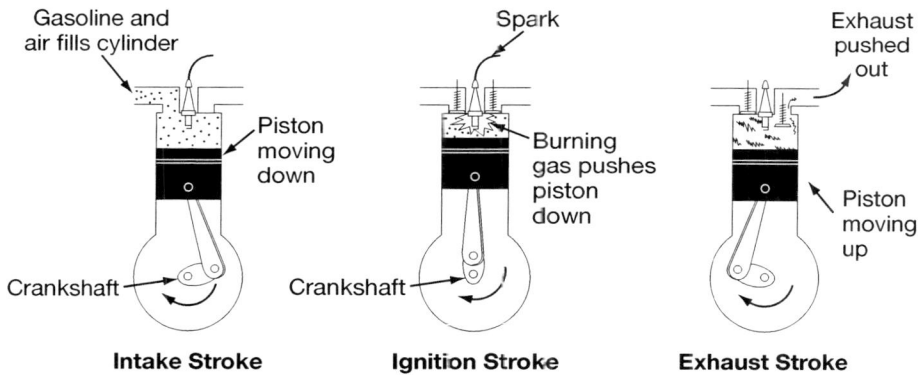

A steam turbine works on the same physical principles, but through a slightly different process. A **steam turbine** is another kind of heat engine. In a steam turbine, heat is converted into movement by the expansion of steam. Water is pumped into a reservoir, where heat is applied. As the water boils, the steam expands and pushes on a turbine. As the steam pushes the turbine, it does work. This causes the temperature of the steam to decrease. The steam then condenses into water again. This water is carried back into the initial reservoir, and the cycle repeats. Again, the gas (steam) gives up some of its heat energy to do work as it pushes the turbine. Steam power provides us with most of our electricity. It is used in all kinds of power stations to drive electricity generators.

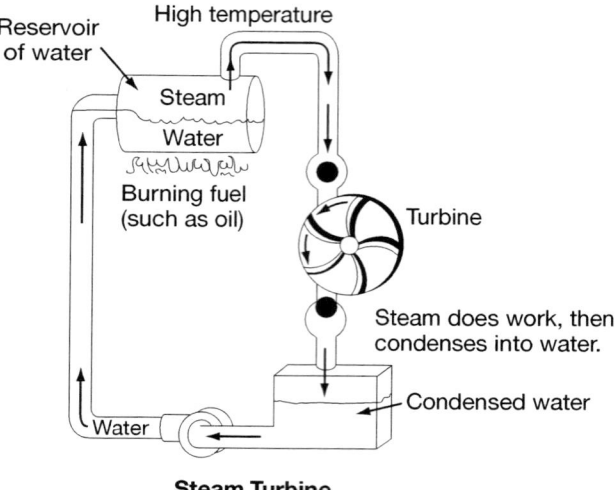

Steam Turbine

All heat engines use some of their heat energy to perform work. In the next section, you will see that there is a limit to how much of the heat energy can actually be converted into the mechanical energy of motion.

Practice 1—Heat Engines

Directions: Decide whether each sentence below describes adiabatic expansion (**E**) or adiabatic compression (**C**). Write the correct letter in each blank.

_____ 1. A bicycle pump feels hot after you use it.

_____ 2. When air rises over a mountain, it becomes cooler than the air in a valley.

_____ 3. When gas performs work in pushing a piston, its temperature decreases.

Directions: Decide whether each statement that follows is true (**T**) or false (**F**). Write the correct letter in each blank.

_____ 4. Heat cannot be converted into mechanical energy.

_____ 5. When the temperature increases, pressure increases because the molecules are moving faster.

_____ 6. When a gas does work pushing against something, its temperature decreases.

_____ 7. In order for an internal combustion engine to function, the gas in the cylinder must be released as exhaust after it pushes the piston.

_____ 8. In a steam engine, hot steam expands and pushes on a turbine, causing the temperature of the steam to increase.

The Laws of Thermodynamics

In the early 1800's, soon after the heat engine was invented, physicists began to study how to design the most efficient engine. In the process, they developed the laws of thermodynamics. **Thermodynamics** is the study of heat and motion. The laws of thermodynamics limit the performance of heat engines.

The first limitation to heat engines is a result of conservation of energy. When heat is added to an engine, the heat changes into other forms. One form is the kinetic energy of the gas molecules—the temperature. Another form is the work done by the heat engine. Because the energy afterward must be the same as the energy before, **the mechanical work done by the engine cannot be greater than the heat energy that is supplied.** This is the **first law of thermodynamics.**

One example of this law is that a car cannot expend any more energy than is in the gasoline in its tank. Gasoline contains about 160 million joules of energy per gallon. Even if all this energy were converted into the motion of the car, this energy would allow an average car to overcome air resistance for only about 200 miles. So, even if a perfectly aerodynamic car were designed, it could get only about 200 miles per gallon. This is due to conservation of energy.

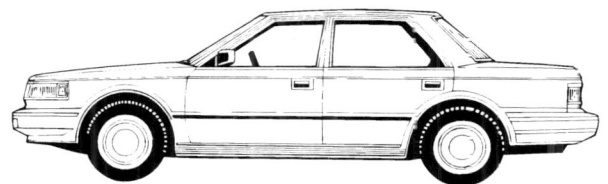

In 1824, however, a French engineer named Sadi Carnot made two discoveries about heat engines. First, Carnot discovered that it is impossible to get out *more* energy from a heat engine than you put into it. Carnot also discovered that it is physically impossible to convert *all* the heat energy from an engine into mechanical motion. Some of the heat energy will be wasted.

In all heat engines, the heat that is added causes a gas to perform some work. Then the gas cools off. The more the gas cools off, the more work it performs. However, unless the gas cools off to absolute zero, it will not give up all of its energy. When the gas is released from the engine, it takes some of this energy away with it.

For example, in a car engine, the burning gasoline creates heat. It does work in pushing the piston, and cools off somewhat. The gas is then released as exhaust. However, the exhaust carries some heat energy away with it.

Carnot calculated the maximum percentage of heat energy that can be converted into mechanical energy or motion. This maximum efficiency is now known as the **Carnot efficiency.** He found that the larger the difference in temperature between the hottest point and coolest point, the more efficient the engine. For example, a steam engine that heats water to 125°C and lets it condense to 25°C, has a maximum efficiency of 25%. No matter how well the engine is designed, it will never convert more than 25% of the heat into energy. The rest of the heat is wasted.

However, increasing the maximum temperature of the gas to 325°C increases the efficiency to 50%. This is why your car may get poor gas mileage in the winter, especially if you don't let it warm up properly. When the engine runs too cold, the difference in temperature between the hot engine and the exhaust is too small. This makes the car not very efficient at converting the heat of burning gasoline into the motion of the car.

The Carnot efficiency is the maximum efficiency a certain type of engine can have. If there is friction or other problems in the engine, it will be less efficient than this theoretical maximum.

Cars could be much more efficient, and get much better gasoline mileage, if the engine temperature were hotter. However, the temperature of the engine is limited by the melting temperature of the engine parts. In the future, cars made of new materials may be made more efficient by running at higher temperatures.

So, a heat engine takes heat energy in, converts some of it to work, and releases some of it as wasted heat energy. **You can't convert all of the input heat into mechanical energy.** This is known as the **second law of thermodynamics.**

The second law is important for engineers, because it tells them that no matter how well an engine is designed, only some of the input energy can be converted into mechanical work. By calculating the Carnot efficiency, they can find the maximum amount. Then, they can try to design a machine that will come as close as possible to the ideal limit. You will learn more about the second law in the next section.

Practice 2—The Laws of Thermodynamics

Directions: Circle the answer that correctly completes each of the following statements.

1. The first law of thermodynamics is a consequence of ____.

    (a) conservation of mass

    (b) conservation of energy

    (c) Carnot efficiency

2. The amount of mechanical work produced by a heat engine is always ____ the heat energy put into the engine.

    (a) less than

    (b) greater than

    (c) equal to

3. A greater difference between the high temperature part of the engine anc the cold temperature part makes the engine ____ efficient.

    (a) more

    (b) less

    (c) equally

4. In a heat engine, ____ of the input heat is converted to mechanical energy.

    (a) all

    (b) some

    (c) none

Directions: Decide whether each statement that follows is true (**T**) or false (**F**). Write the correct letter in each blank.

____ 5. A steam engine that heats the steam to 500°C is less efficient at creating motion than an engine that only heats the steam to 200°C.

____ 6. The exhaust from your car carries away some of the input heat.

Entropy

The **second law of thermodynamics** says that you can never convert all of the heat into mechanical energy or motion. This is an example of a more general principle called entropy. **Entropy** is the amount of disorder in a system. The entropy principle states that all systems tend to go from a state of order to a state of disorder.

As an example, suppose you have a box full of coins. All of the coins are in straight rows facing "heads" up. The coins are in an ordered state. If you start to shake the box, the coins will become more and more disorganized. These coins now have more entropy than the unshaken coins. No amount of shaking will bring them back to their original ordered state.

Now, imagine a jar full of gas molecules. Because the molecules are contained in a jar, they are in an ordered state. If you open the lid, the molecules will drift out of the bottle, spreading throughout the room in a disorganized state. No matter how long you wait, the molecules will not go back into the bottle.

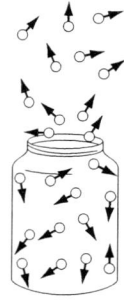

Closed jar—
gas molecules
are trapped

Open jar—
gas molecules
drift out

Similarly, mechanical energy, such as the kinetic energy of a rolling wheel or the potential energy of a stretched bow and arrow, represents an ordered state. The mechanical energy can be used for something, such as moving a car or launching an arrow. Eventually, that mechanical energy will be converted into heat by friction or air resistance. The wheel will slow down, or the arrow will stick into a tree. Now, the useful mechanical energy has been converted into heat, which is the random motion of molecules. And, the second law of thermodynamics says that it is physically impossible to convert all of that heat back into useful mechanical energy.

There are many examples of useful energy being converted to nonuseful energy. When you leave the lights on in your house, the useful electrical energy is converted to light. Eventually, the light is converted to heat when the light is absorbed by the surroundings. When gasoline is burned in your car, some of the useful chemical energy is converted into the motion of the car. But, some is always (according to the second law) converted into heat in the exhaust. Eventually, the car will stop. So, even the mechanical energy of its motion will eventually be transformed to heat by friction in the brakes. Heat energy has more entropy than mechanical energy. The second law of thermodynamics implies that **for natural processes, entropy always increases.**

Of course, you can use energy to undo the disorder in a system. In our example of the jar on page 95, you can use an air pump to push the molecules of gas back into the jar. However, the energy spent in running the pump will always produce more disorder than you eliminated. Overall, the entropy increases. If you reduce the entropy in one place (inside the jar), you increase it by even more somewhere else (outside the jar).

Entropy has a negative effect on the environment. Power plants, factories, and automobile engines use heat energy to create mechanical energy. However, the second law says that they will always release some of this heat energy as a by-product. Thermal pollution disturbs plants and animals by changing the temperature at which they must live. And, because you cannot decrease entropy, you cannot get rid of thermal pollution once it is produced. It is an inevitable consequence of the laws of thermodynamics. However, by reducing the amount of energy we consume, we can reduce the amount of thermal pollution that is created in the process.

Practice 3—Entropy

Directions: Decide whether each statement that follows is true (**T**) or false (**F**). Write the correct letter in each blank.

_____ 1. Nonuseful forms of energy, such as heat, can be completely converted back into mechanical energy.

_____ 2. When gas escapes from a jar, entropy decreases.

_____ 3. The amount of entropy in the universe never changes.

_____ 4. When you leave the lights on in your house, electrical energy is converted to nonuseful energy, increasing entropy.

_____ 5. It is possible to use energy to reduce the entropy of one object, although the entropy of something else must always increase.

■ LESSON MASTERY TEST

Directions: Circle the answer that correctly completes each of the following statements.

1. _____ describes the maximum percentage of input energy that can be converted into mechanical energy by an engine.

   (a) Specific heat

   (b) Carnot efficiency

   (c) Thermal equilibrium

2. When a gas pushes against something to expand, its temperature _____.

   (a) increases

   (b) decreases

   (c) remains the same

3. Entropy is a measure of the _____ of a system.

   (a) order

   (b) disorder

   (c) reorder

4. In a heat engine, _____ of the input heat is converted to mechanical energy.

   (a) all

   (b) none

   (c) some

Directions: Decide whether each statement that follows is true (**T**) or false (**F**). Write the correct letter in each blank.

_____ 5. The amount of entropy in the universe never changes.

_____ 6. When you push on a gas to compress it, its temperature increases.

_____ 7. An engine will never be as efficient as the Carnot efficiency predicts, because friction and other losses reduce the efficiency.

_____ 8. Nonuseful forms of energy, such as heat, can be completely converted back into mechanical energy.

PART 3

SOUND AND LIGHT

Table of Contents

■ Lesson 1—Sound Waves

Goals: To understand the properties of waves; to learn how the properties of waves apply to sound

Describing Waves

Any repeated back-and-forth motion is called a **vibration.** For example, swinging a pendulum back and forth, or wiggling your hand, are both vibrations.

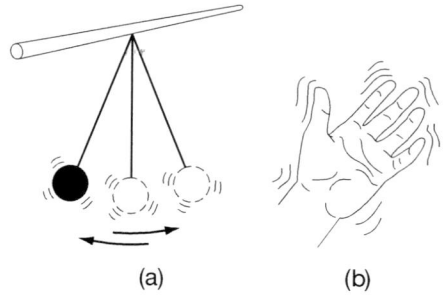

When a vibration moves from one place to another, it is called a **wave.** In the diagram below, the vibration of the hand creates a wave that travels along the rope. When you drop a rock into water, the vibrations in the water spread out across the surface.

(a)   (b)

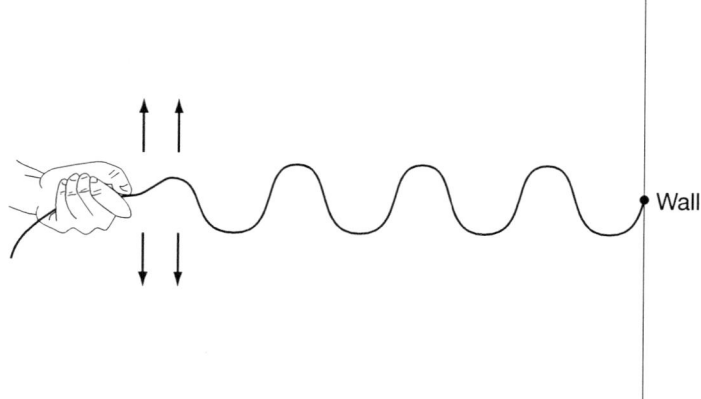

Wall

An important property of waves is that the vibration travels *without* any material moving along with it. When you throw a ball, the ball itself moves from one place to another. In the diagram above, when the person shakes the rope, the "wiggle" will move down the rope to the other end. However, the whole rope did not move from one place to another, as happens when you throw a ball. Only the vibration moved. Similarly, when you speak, the air that comes out of your mouth does not travel into other people's ears. Only the vibration of the air molecules is transmitted.

The substance that a wave travels through is called the **medium.** In the previous diagram, the medium is the rope. For ocean waves, the medium is water. For sound waves, the medium is air. In a wave, the medium does not move from one place to another. Only the energy or disturbance caused by the vibration is transmitted.

There are many terms that we use to describe waves and vibrations. Several of these are illustrated in the diagram below.

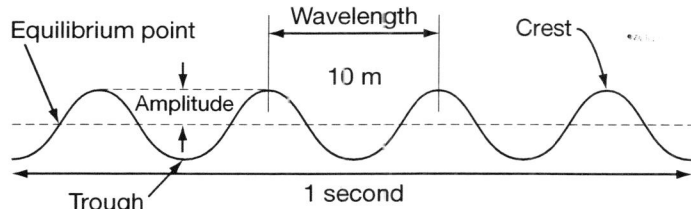

The top of a wave is known as a **crest.** The bottom of a wave is known as a **trough.** The resting position, in between the crest and trough (the dashed line in the art above), is known as the **equilibrium point.**

The **amplitude** of a wave describes how big it is. The amplitude is measured as the distance from the equilibrium point to the crest of the wave. The taller the wave, the greater its amplitude.

The **wavelength** describes how long each wave is. The wavelength is usually measured as the distance from one crest to the next. Or, it can be measured as the distance from one trough to the next. It can also be the distance between any two identical points in the wave.

Wavelength and amplitude are both distances. So, they are measured in the same units as length. In the metric system, they can be measured in meters, centimeters, or even nanometers, depending on how big or small the wave is.

We can also measure how quickly waves or vibrations oscillate, or move back and forth. There are two terms to describe this. The **frequency** is the number of complete waves or vibrations that occur in a certain amount of time. High frequency means the waves are occurring very quickly. Low frequency means the waves are being produced infrequently. For example, if you notice that 5 waves hit the beach each minute, the frequency is 5 waves per minute.

©1998, 2001 J. Weston Walch, Publisher

The most common unit used to measure frequency is the **hertz (Hz).** Hertz was named after the German scientist, Heinrich Hertz. Hertz is the number of waves per second. For example, if you are holding a rope and you shake your hand up and down 3 times each second, the waves in the rope will have a frequency of 3 Hz.

Higher frequencies are usually expressed as **kilohertz (kHz)** or **megahertz (MHz).** One kilohertz is one thousand waves per second. One megahertz is one million waves per second.

The frequency of a radio station is simply the frequency of the radio waves that are carrying the signal. For instance, FM radio stations are in the range of 88 MHz to 108 MHz. If you are listening to 104 FM, it means the radio waves are vibrating 104 million times each second. On the other hand, AM stations are in the kHz range. If you listen to 950 AM, it means the waves are vibrating at 950 kHz, or 950 thousand times each second.

You can also measure how quickly waves are oscillating by the period. The **period** is the length of time it takes for one wave to complete. For example, one wave hits the beach every ten seconds. This means that the period is 10 seconds. If it takes you 0.2 seconds to shake your hand up and down once, then the period of the waves you will create is 0.2 seconds.

There is a simple relationship between frequency and period. They are inverses, or opposites. If a wave has a high frequency, then each wave does not take much time. So, it has a short period. If each wave takes a lot of time (long period), then there will not be many of them each second. So, the frequency will be low.

This can be summarized in the equations below.

frequency = 1 / period

period = 1 / frequency

For example, if the frequency of a wave is 5 Hz (5 waves each second), then each wave takes ⅕ second, making the period ⅕ second.

Practice 1—Describing Waves

Directions: Circle the correct answer to each of the following questions.

1. What term describes the substance a wave travels through?

    (a) vibration   (b) medium   (c) matter

2. What is the top of a wave called?

    (a) crest   (b) trough   (c) equilibrium point

3. If a wave oscillates twice each second, what is its period?

    (a) 2 seconds   (b) 2 Hz   (c) ½ second

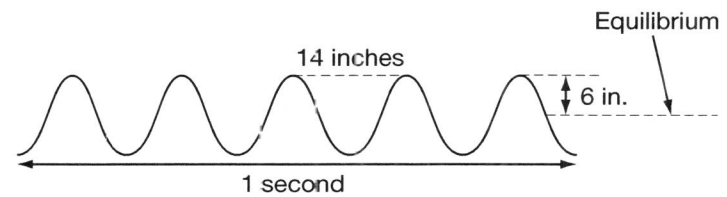

4. What is the amplitude of the wave above?

    (a) 14 inches   (b) 12 inches   (c) 6 inches

5. What is the wavelength of the wave above?

    (a) 14 inches   (b) 7 inches   (c) 6 inches

6. What is the frequency of the wave above?

    (a) 14 Hz   (b) 5 Hz   (c) 1 Hz

Motion of Waves

In the diagram of the rope on page 100, the rope is vibrating up and down. But, the waves are traveling horizontally. Likewise, water waves wiggle up and down. But, the waves spread out across the surface. These are known as transverse waves. In a **transverse wave,** the medium vibrates in a different direction than the waves travel.

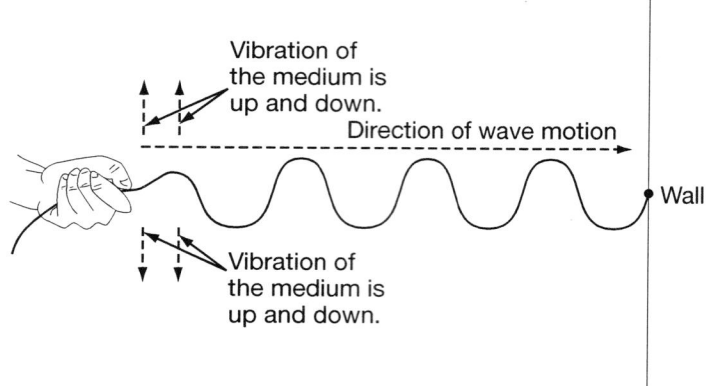

Vibration of the medium is up and down.

Direction of wave motion

Wall

Vibration of the medium is up and down.

On the other hand, if you push a coiled spring back and forth, as shown below, the vibrations are horizontal. And, the wave is traveling horizontally also. The vibration of the medium is in the *same* direction as the wave is traveling. This is known as a **longitudinal wave.**

You have already learned about frequency and period, two terms that describe how quickly waves are produced. We can also describe how quickly a wave travels from one place to another. This is the wave's **speed.** Ocean waves that are moving quickly have a high speed. Slow waves on a pond have a low speed. Just like the speed of any other object, a wave's speed is the distance a wave moves *per time.* For example, if a wave moves 5

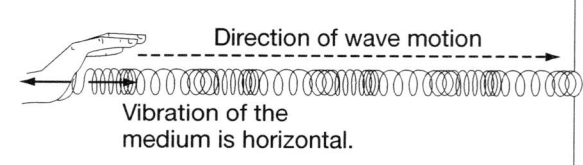

Direction of wave motion

Vibration of the medium is horizontal.

meters in one second, then its speed is 5 m/sec. If you are riding in a boat that is traveling at a rate of 20 miles per hour, and the waves are traveling at the same rate as the boat, then the speed of the waves is 20 miles per hour.

Tip!

It is important to remember the difference between speed and frequency. Frequency is how quickly waves oscillate. If you are watching waves at the beach, the frequency is the number that hit the shore in a certain amount of time. Speed is how quickly the waves move from one place to another. At the beach, the speed of a wave is how fast it moves from a distant point in the ocean to shore.

The speed at which waves move generally depends only on the medium it is moving in. (You will learn later that light is an exception to this rule.) The speed does not depend on the frequency, wavelength, or amplitude of the waves. For example, large water waves and small water waves travel at the same speed. High-frequency sounds travel at the same speed as low-frequency sounds. And, if you shake the rope shown below faster, you will produce more waves. But, the waves will not reach the other end of the rope any faster.

The only way to change the speed that waves travel in a medium is to change something about the medium. For instance, if you stretch the rope tighter, the waves will travel more quickly. Air that is heated transmits sound more quickly than cooler air. This is because the faster moving molecules collide more often, transmitting the sound more quickly.

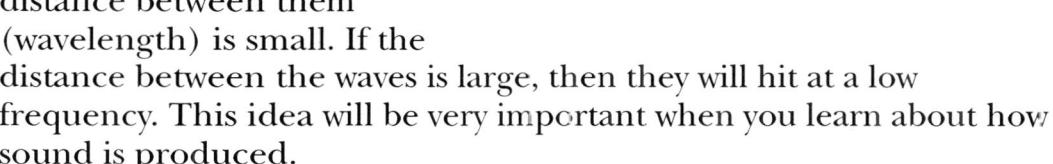

Shaking the rope faster will not change the speed of the wave. However, it will change the wavelength. As the frequency of the shaking increases, the wavelength becomes shorter. If you shake the rope more slowly, the wavelength becomes longer.

Similarly, if the ocean waves hit the beach often, then the distance between them (wavelength) is small. If the distance between the waves is large, then they will hit at a low frequency. This idea will be very important when you learn about how sound is produced.

When a wave passes through a narrow opening, it can spread out on the other side. You may have noticed this in water waves. This effect is known as **diffraction.** The narrower the opening, the more the waves spread out on the other side.

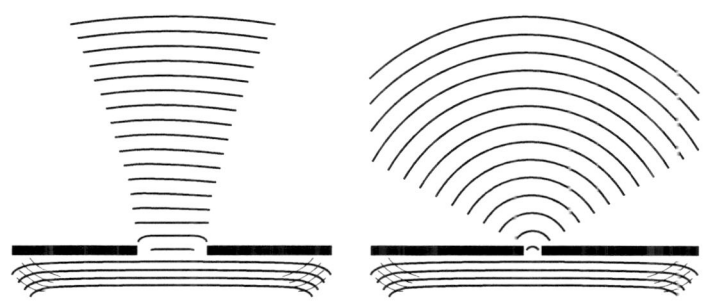

Suppose two people hold opposite ends of a rope. They both shake the rope. As the waves travel down the rope, they will overlap. The point at which the waves overlap is known as **interference.** Similarly, if you drop two rocks at different places in a pond, the waves produced by each rock will overlap. This creates interference.

When two crests overlap each other, they add up to make an even bigger crest than before. This is known as **constructive interference.** The diagram at the right illustrates this.

The same thing would happen if two troughs overlapped. They would make an even larger trough. When two waves are "in step" like this, we say that they are **in phase.** After the waves have passed through each other, they continue traveling normally.

On the other hand, if a crest and a trough overlap, they will momentarily cancel each other out. This is known as **destructive interference.** The waves are not actually destroyed by this, however. They continue moving through each other. Once they no longer overlap, they travel as they did before. Two waves in opposite motion like this are called **out of phase.**

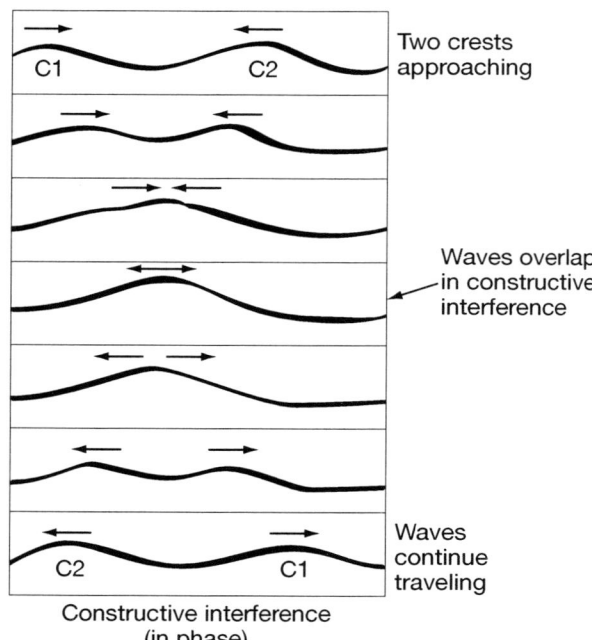

Two crests approaching

Waves overlap in constructive interference

Waves continue traveling

Constructive interference
(in phase)

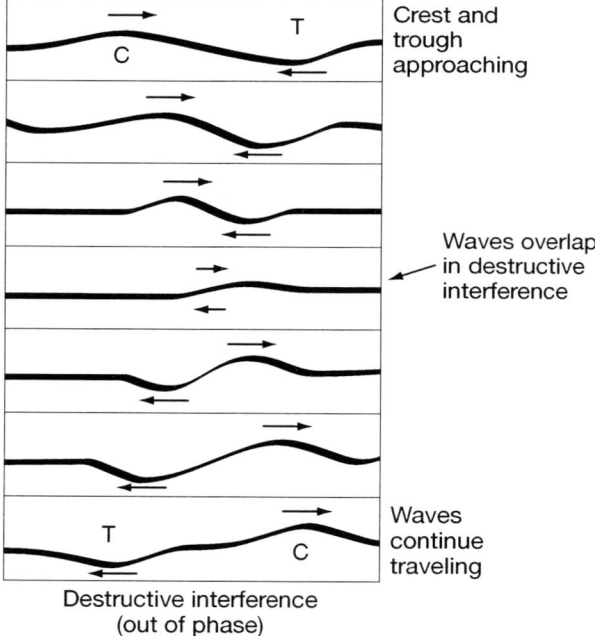

Crest and trough approaching

Waves overlap in destructive interference

Waves continue traveling

Destructive interference
(out of phase)

Remember that when two waves meet each other, they are not permanently changed. They undergo interference as they overlap. But, then they continue to travel as they did before hitting each other.

Practice 2—Motion of Waves

Directions: Decide whether each statement that follows is true (**T**) or false (**F**). Write the correct letter in each blank.

_____ 1. In a transverse wave, the medium vibrates in the same direction that the wave moves.

_____ 2. Ocean waves are transverse waves.

_____ 3. A wave with a high frequency vibrates very quickly.

_____ 4. By shaking a rope faster, you can make the waves move down the rope more quickly.

_____ 5. In order to change the speed of a wave, you must change something about the medium.

_____ 6. Diffraction is the process where waves spread out as they pass through an opening.

_____ 7. The larger the opening a wave passes through, the more it will diffract.

_____ 8. Waves that are matched crest to crest and trough to trough are called "out of phase."

_____ 9. When two waves interfere destructively, they stop moving completely.

Sound Waves

Sound is a longitudinal wave in air. Sound is produced by vibrating objects. Just as the motion of a hand creates waves in a rope, the back-and-forth vibration of the speaker creates waves in the air.

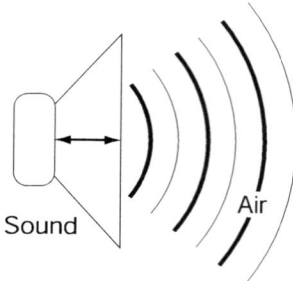

As the surface of the speaker pushes forward, it pushes the air molecules closer together. The area where the molecules are pushed together is called a **compression.** These molecules push on the ones in front of them, causing them to move. Then the whole disturbance moves forward through the air. This is similar to when the person at the back of a line pushes on the person in front of him, who pushes on the next person, and so on.

After the surface of the speaker pushes forward, the surface pulls back. This creates a rarefaction. A **rarefaction** is an area with few air molecules. The sound wave travels out as a series of compressions and rarefactions. The compressions are the crest of the wave. The rarefactions are the troughs of the wave.

The wavelength of the sound wave is the distance between two compressions. The frequency is the number of times the speaker vibrates each second. The amplitude is determined by how much the molecules are pushed together in each compression. The more the molecules are pushed together, the greater the amplitude.

Any vibrating object, such as a plucked guitar string or the buzzing of your vocal cords, creates sound waves like this.

When the sound wave enters your ear, it causes your eardrum to begin vibrating. Your eardrum vibrates with the same frequency and amplitude as the sound wave. Your brain then detects this as the sound that you hear.

The **loudness** of a sound depends on the amplitude. The more the molecules are pushed together in the sound wave, the harder they will push on the eardrum. This is what creates a louder sound.

The loudness of a sound is measured in **decibels (dB).** Zero decibels is the quietest sound the normal ear can hear. Each ten decibel increase above this means that the sound is 10 *times* louder. So, 10 dB is 10 times as loud as 0 decibels, 20 dB is $10 \times 10 = 100$ times as loud, 30 dB is $10 \times 10 \times 10 = 1,000$ times as loud, and so on. Likewise, a 60 dB sound is 10 times as loud as a 50 dB sound, and 100 times as loud as a 40 dB sound.

The following table shows the decibel values of some common sounds.

Acoustic Condition	Decibels (dB)
Threshold of hearing	0
Bedroom at night	30
Conversation at one yard	60
Subway train	90
Rock band	110
Threshold of pain	140
Saturn rocket at close range	200

The **pitch** of a sound wave (how high or low the sound wave is) depends on the frequency. A high-frequency sound wave has a high pitch, like a shrill whistle. A low-frequency sound wave has a low pitch, like a bass drum.

The human ear can detect frequencies between 20 Hz and 20 kHz. In other words, objects vibrating between 20 times per second and 20,000 times per second create sounds that the ear can hear. Sounds that are below this range are called **infrasonic.** If you wave your hand back and forth 5 times per second, you are creating an infrasonic sound wave. Sound waves whose frequencies are too high to be heard are called **ultrasonic.**

In Real Life

Ultrasonic sound waves are often used in medicine. One way doctors use these waves is to look at the inside of the body. Ultrasonic waves are sent into the body, and reflect off the internal organs. The reflection (an ultrasonic echo) is used to make an image of the internal organs. This is known as an ultrasound. It is often used to look at an unborn baby in the womb. Ultrasonic waves can also be used to remove kidney stones. The sound waves are focused on the stone, and the vibration created in the stone causes it to break apart. It can then be expelled by the body.

The ultrasonic waves in an ultrasound are not dangerous radiation. They are simply regular sound waves whose frequency is too high to be heard. They are produced by a speaker, just like any other sound.

Like the speed of other waves, the speed of sound can be changed only by changing the medium. At room temperature, sound waves travel at 340 m/sec, or about 750 mph. This is true for sounds that are loud or soft, or sounds that have a high pitch or low pitch. If the air is warmer, the speed will increase. In fact, sound travels about 5 mph faster for each 10°F increase in temperature. Also, sound can travel through materials other than air. For instance, when you are under water, you still hear sound. This is because the vibrations travel through the water by compressions and rarefactions of the water molecules. Sound travels about four times faster in water than in air.

Sound can also travel through solids. If you put your ear to the desk and tap on it, you will hear the sound waves transmitted through the desk. Sound travels even faster through solids than through liquids or gases. This is because the tight bonds between the molecules allow them to transmit the vibrations easily.

Like other waves, sound waves can reflect off objects they hit. When you hear an echo in the mountains, you are simply hearing the sound wave reflecting off a distant mountainside. Because sound only travels about 1,000 feet each second, it would take 2 seconds for the sound to go out and come back from a mountain 1,000 feet away. Therefore, you would hear the echo 2 seconds later.

Sound reflects much better off hard, smooth surfaces than off soft, rough surfaces. That is why large cliffs make such good echoes. However, echoes can ruin the sound quality in concert halls or auditoriums. So, the walls of these rooms are often covered with a material that will absorb the sound, rather than reflect it.

Practice 3—Sound Waves

Directions: Decide whether each statement that follows is true (**T**) or false (**F**). Write the correct letter in each blank.

_____ 1. All sound is created by a vibrating object pushing on the air.

_____ 2. The region where air is pushed closer together is called a rarefaction.

_____ 3. A 50 dB sound is 100 times as loud as a 30 dB sound.

_____ 4. An object that is vibrating very rapidly will produce a higher pitched sound.

_____ 5. The human ear can detect all frequencies of sound.

_____ 6. Sounds with a frequency greater than 20,000 Hz are called ultrasonic.

_____ 7. Sound travels more quickly in solids than in air because the tight bonds between the molecules allow them to transmit the vibrations easily.

_____ 8. Sound travels more slowly through hot air, because hot air is denser.

_____ 9. An echo is produced whenever sound is absorbed.

Making Sound

All sound is created by vibrating objects. However, what determines the frequency and amplitude of the sound that will be created?

Amplitude is determined by two things: how much the source of the sound is vibrating, and how well it transmits the vibrations into the air. For example, the harder you pluck a guitar string, the larger the amplitude of its vibrations will be. The larger the amplitude, the louder the sound. Likewise, a large speaker is better at transmitting its vibrations into the air because it pushes on more air molecules. This will make a louder sound.

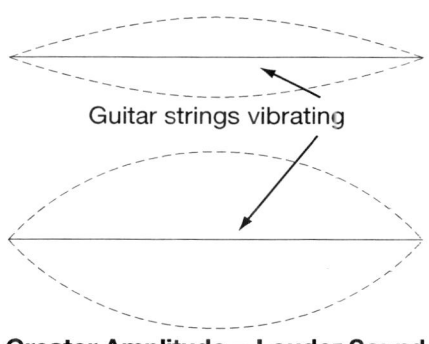

Guitar strings vibrating

Greater Amplitude = Louder Sound

The frequency of the sound wave depends on the frequency of the object that created it. If you tap a fork on the table, it will always ring at the same frequency. If you hit an empty glass, it will make the same sound. We call this the **natural frequency** of the object.

The natural frequency of an object depends on many factors. These factors are the elasticity of its material, its shape, and its size. You can change the natural frequency by changing any of these factors.

Try tapping a glass with a fork while you gradually fill the glass with water. As it fills, the sound will become lower in pitch. This means that the natural frequency of the glass decreased. This is because the mass of the glass increased. With more mass, the glass cannot vibrate back and forth as quickly. So, the natural frequency becomes lower.

When you speak, your vocal cords are vibrating. Their natural frequency is determined by the muscles that are attached to them. These muscles can stretch the vocal cords. When the vocal cords are stretched, the natural frequency is higher. Therefore, your voice will have a higher pitch.

Musical instruments make use of natural frequency to play certain notes. For example, the natural frequency of a guitar string is determined by its length. Remember that in any medium, high-frequency waves will have a short wavelength, while low-frequency waves will have a long wavelength. Thus, a long guitar string will have a longer wavelength. Therefore, it will have a lower frequency. By pressing your finger on the string, you shorten the wavelength and increase the frequency. This creates a sound with a higher pitch.

When you tune the guitar string, you are adjusting the tension in the string. By changing the medium (the string), you change the frequency of the waves. More tension in the string causes it to snap back faster when plucked. This makes a higher natural frequency.

The different strings on the guitar also have different thicknesses. A thicker string has more mass, so it doesn't accelerate as quickly. This creates a lower natural frequency because the string doesn't vibrate as quickly.

Plucking the string harder *will not* change the pitch of the sound, however. This changes the amplitude, but not the natural frequency. The natural frequency is determined by the characteristics of the object creating the sound.

If you look inside a piano, you will notice that some strings are long and thick, while others are short and thin. Which do you think correspond to the high notes?

Most objects don't just vibrate at exactly one frequency. For example, a guitar string can vibrate as a whole, or it can vibrate in halves or thirds or fourths, as shown below.

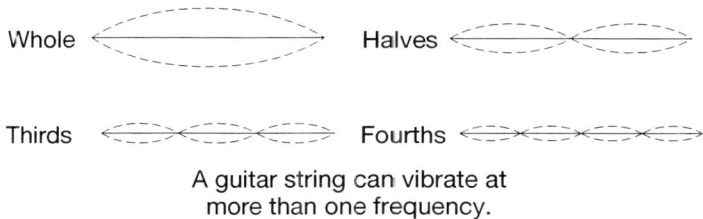

A guitar string can vibrate at
more than one frequency.

These extra vibrations create additional frequencies, called **overtones.** A pure sound has no overtones. But, a very rich sound has many overtones. Even at the same pitch, different instruments have different sounds. This is because they produce different overtones. For example, a flute makes a very pure sound because it creates very few overtones. But, a saxophone produces many overtones.

Practice 4—Making Sound

Directions: Circle the answer that correctly completes each of the following statements.

1. Turning up the volume on your stereo increases the ____ of the sound waves.

    (a) wavelength   (b) speed   (c) amplitude

2. The natural frequency of an object is NOT affected by ____.

    (a) size   (b) elasticity   (c) color

3. Increasing the mass of an object generally ____ its natural frequency.

    (a) increases

    (b) decreases

    (c) does not change

4. Shortening a guitar string decreases its ____, and increases its ____.

    (a) frequency, wavelength

    (b) wavelength, frequency

    (c) wavelength, amplitude

5. Increasing the tension in a guitar string ____ its natural frequency.

    (a) increases   (b) decreases   (c) does not affect

6. The additional frequencies at which an object vibrates are known as its ____.

    (a) overtones

    (b) natural frequencies

    (c) wavelengths

The Doppler Effect

If you stand still as a train passes you, you will notice differences in sound. As the train approaches, its engine sounds higher in pitch than when it is going away from you. Or, as a fire truck passes you, its siren goes from high pitch to low pitch. This is known as the **Doppler effect.**

Sound is a wave traveling in the air, and the frequency of the wave determines its pitch. As the siren comes toward you, the sound waves in front of it become closer together. This is because during the time between two vibrations, the fire truck moves forward a little, "catching up" with the sound wave. When these sound waves reach you, they are closer together, so you hear a higher frequency. This makes the pitch higher.

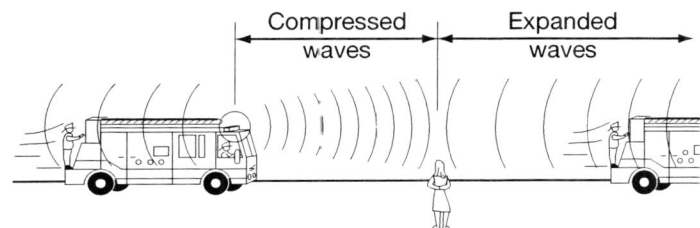

Compressed waves   Expanded waves

The Doppler Effect

Likewise, on the back side of the fire truck, the waves become farther apart, or stretched out. This is because the fire truck moves away from each consecutive sound wave. When the fire truck is moving away from you, you hear these stretched-out waves. They have a lower frequency. So, the pitch is lower.

Tip!

Remember: As the source of the sound is moving toward you, the frequency is higher, because the waves are closer together. As the source is moving away from you, the frequency is lower, because the waves are farther apart.

The faster an object moves, the more the sound waves in front of it will be compressed. If the object is moving at the speed of sound or faster, the waves will actually all pile up on top of each other, in one pulse of sound. This pulse of sound is known as a **sonic boom.** All of the sound energy is concentrated in one peak. Jets that travel faster than the speed of sound create a sonic boom. When a plane is flying faster than the speed of sound, it is flying faster than the compression waves it makes. The plane compresses those waves. This compression is heard on the ground as a sonic boom. The sound of a bullet is actually a small sonic boom. Even the cracking of a whip is a small sonic boom, as the end of the whip travels faster than the speed of sound.

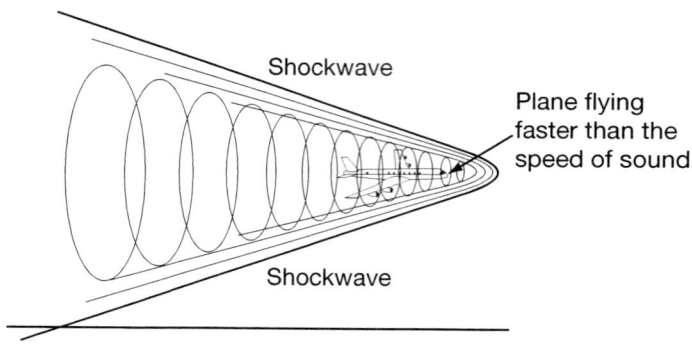

The Doppler effect actually applies to all types of waves, including radar and light waves. Police radar uses the Doppler effect to measure the speed of cars. The radar unit sends a beam of waves at a car, and the car reflects them back. If the car is stationary, the reflected waves will have the same frequency. If the car is moving, however, the reflected waves will have a different frequency, due to the Doppler effect. By measuring the difference in frequency, the police can determine how fast the car is moving.

If an object is moving close to the speed of light, the light from the object will change due to the Doppler effect. Light from an object moving away will be lower in frequency, closer to the red end of the light spectrum. We say it is **red-shifted.** About 50 years ago, astronomers noticed that light from many stars is red-shifted. They concluded that distant stars are moving away from us. In fact, the entire universe is expanding.

Thus, in addition to explaining why moving objects sound different, the Doppler effect can be used to measure moving objects, from speeding cars to distant galaxies.

Practice 5—The Doppler Effect

Directions: Circle the answer that correctly completes each of the following statements.

1. The Doppler effect explains why the _____ of a moving object's sound changes.

   (a) amplitude

   (b) frequency

   (c) mass

2. As a train moves towards you, the frequency of its sound _____.

   (a) becomes higher

   (b) becomes lower

   (c) stays the same

3. When an object moves faster than the speed of sound, the crests of all the waves overlap, producing a _____.

   (a) destructive interference

   (b) red-shift

   (c) sonic boom

4. Light from stars that are moving away from us gets shifted to a _____ frequency, towards the _____ end of the spectrum.

   (a) higher, red

   (b) lower, blue

   (c) lower, red

■ LESSON MASTERY TEST

Directions: Using the diagram below, circle the correct answer to each of the following questions.

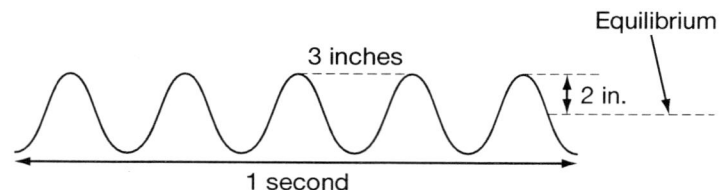

1. What is the amplitude of the wave?

   (a) 2 inches   (b) 3 inches   (c) 4 inches

2. What is the wavelength of the wave?

   (a) 2 inches   (b) 3 inches   (c) 4 inches

3. What is the frequency of the wave?

   (a) ⅕ Hz   (b) 3 Hz   (c) 5 Hz

4. What is the period of the wave?

   (a) ⅕ second   (b) ⅓ second   (c) 5 seconds

Directions: Circle the correct answer to each of the following questions.

5. What term describes the substance a wave travels through?

   (a) vibration   (b) medium   (c) matter

6. What is the bottom of a wave called?

   (a) crest   (b) trough   (c) equilibrium point

7. If a wave oscillates twice each second, what is its period?

   (a) 2 seconds   (b) 2 Hz   (c) ½ second

8. What term describes a wave spreading out as it passes through a narrow opening?

   (a) interference   (b) diffraction   (c) vibration

(continued on next page)

9. What unit is used to measure the loudness of a sound?

    (a) Hertz   (b) decibels   (c) meters

10. What are the additional frequencies called at which an object vibrates?

    (a) overtones

    (b) natural frequencies

    (c) wavelengths

Directions: Decide whether each statement that follows is true (**T**) or false (**F**).
Write the correct letter in each blank.

_____ 11. A 50 dB sound is 100 times as loud as a 30 dB sound.

_____ 12. An object that is vibrating very rapidly will produce a lower pitched sound.

_____ 13. An echo is produced when sound waves reflect off a smooth surface.

_____ 14. When an object is moving toward you, the frequency of the sound you hear becomes higher.

_____ 15. Increasing the mass of a vibrating object increases its natural frequency.

_____ 16. Waves that are matched crest to crest and trough to trough are called "out of phase."

©1998, 2001 J. Weston Walch, Publisher

■ Lesson 2—The Nature of Light

Goal: To understand the properties of light as both waves and particles

Light Waves

Everything that we see is the result of light waves. They reflect off objects around us into our eyes. However, the true nature of light is still not completely understood, even by physicists.

Light displays many of the properties of waves, including radio waves, X-rays, and microwaves. All of these waves are known together as **electromagnetic radiation.** The different forms of electromagnetic radiation have different frequencies. Radio waves have low frequencies, and X-rays have very high frequencies. Light is in the middle.

Remember that low-frequency waves have a long wavelength, while high-frequency waves have a short wavelength. Thus, radio waves have long wavelengths of up to 100 meters. X-rays have a wavelength less than one-billionth of a meter.

Similarly, different colors of light have different frequencies. Blue light has a high frequency, and red light has a low frequency. The wavelength of visible light ranges between 0.5 millionths of a meter for blue light to 0.8 millionths of a meter for red light. Only electromagnetic radiation in this tiny range can be detected by human eyes.

For many years, scientists couldn't figure out what the medium was for electromagnetic radiation. They thought that light must have a medium since sound waves travel through air, and ocean waves travel through water. However, light can travel through empty space. The light from the sun travels through millions of miles of empty space. So, unless there was some substance hidden in outer space, there was no medium for light.

About a hundred years ago, scientists came up with the first explanation for how light is created and transmitted. This explanation demonstrated why light has no medium. All waves are created by something that is vibrating. You can create sound waves by vibrating your vocal cords. And, you can create waves in a rope by wiggling your hand.

All forms of electromagnetic radiation, including light, are created by the vibration of charged particles, specifically electrons. Around an electron is an electric field and a magnetic field. The **electric field** is the pattern of its electric force. The **magnetic field** is the pattern of its magnetic force. A wiggling electron causes the electric and magnetic fields around it to change. As the electron moves back and forth, these disturbances spread out around it. This is much like wiggling a stick in a pond and causing disturbances in the water to spread out.

However, these electromagnetic disturbances do not need any medium to carry them. In fact, the electric and magnetic fields *are* the medium. This might seem like a strange concept. It actually took scientists several decades to finally accept this fact.

Consider a radio station that transmits radio waves from its antenna. Inside the antenna, electronic circuits provide a rapidly alternating push to the electrons in the antenna. As the electrons vibrate, they put out electromagnetic waves with the same frequency at which the electrons are alternating. For example, if the electrons alternate 96,000,000 times each second, then the station sends out waves of 96 MHz. That station's frequency is 96 FM.

Electromagnetic waves

Electrons vibrating

Because electromagnetic waves don't require a medium, they can pass through empty space. All electromagnetic waves travel at the same speed through empty space. This speed is amazingly fast—about 180,000 miles per second. That is why light seems to travel almost instantaneously. This is also why telephone and television signals can be sent around the world with almost no delay.

Electromagnetic waves can also pass through matter, such as light going through a window. However, changing the medium changes the speed of the waves, just as the speed of sound changes when traveling through different substances. Traveling through matter always makes light travel slower. Light only travels ¾ as fast through water, and ⅔ as fast through glass.

Practice 1—Light Waves

Directions: Circle the answer that correctly completes each of the following statements.

1. Electromagnetic waves do NOT include ____.

   (a) radio waves   (b) light   (c) sound waves

2. The frequency of light is ____ than the frequency of X-rays.

   (a) higher   (b) lower

3. Waves with a high frequency have ____ wavelength.

   (a) a shorter   (b) a longer   (c) the same

4. The medium of electromagnetic waves is ____.

   (a) air   (b) electrons   (c) electric and magnetic fields

5. Electromagnetic waves are created by the vibration of ____.

   (a) atoms   (b) speakers   (c) electrons

Light: Wave or Particle?

Much of the early evidence showed that light and other electromagnetic radiation are waves. For instance, light can undergo interference. Two light waves that are out of phase cancel each other when they overlap. Two light waves that are in phase create an even brighter wave. Light can diffract, or spread out, when passing through a narrow opening. Furthermore, radio waves and microwaves can be created by wiggling electrons back and forth, just like wiggling a rope.

However, at the beginning of this century, several experiments showed that light could act as a particle as well. For instance, it was shown that light always travels in chunks or bundles, rather than in continuous waves. These bundles of light energy are called **photons.** Photons hitting an object are like baseballs hitting a wall, rather than like waves hitting the shore. In other words, photons of light act like particles.

Furthermore, in 1913 the Danish scientist Niels Bohr proposed a new theory of how light is created. Remember that electrons in an atom orbit around the nucleus. Bohr suggested that the electrons cannot travel in any path they want to. They can travel only in specific paths called **orbits.** Electrons can switch between these paths. But, when they do so, they release some of their energy as a bundle of electromagnetic radiation. Each bundle is one photon.

The farther apart the electron has to jump between orbits, the more energy the photon will have. The more energy an electron has, the higher its frequency will be. Thus, a single photon of blue light has more energy than a single photon of red light. The amount of energy determines the frequency of the photon. Therefore, the amount of energy determines its color.

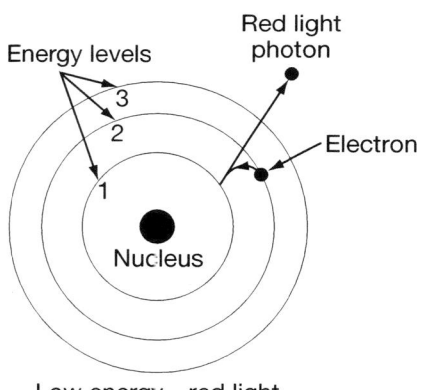

Low energy—red light

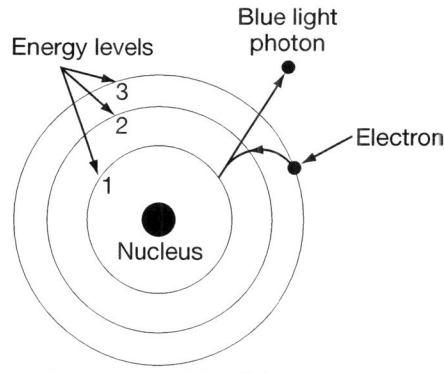

High energy—blue light

What makes an electron jump between these orbits? Usually an electron stays in the closest available orbit to the nucleus. However, several processes can push it into a higher orbit. The electron can collide with other atoms or electrons. Or, another photon can hit the atom. This gives the electron extra energy. Later, the electron can spontaneously, without any outside influence, fall back down into the lower orbit. As the electron falls, it releases the energy it had. This energy is in the form of a photon of light. The farther the electron falls down, the more energy it releases.

In the particle view of light, the color of the light depends on the energy of each individual bundle, or photon, of light. The brightness of the light depends on how many photons there are. More photons mean a brighter light beam.

But, is light an electromagnetic *wave*, or is it a photon *particle?* The answer is both. Sometimes light behaves as a particle, and sometimes it behaves as a wave. Its actual nature is probably something in between, which is still beyond the understanding of scientists. This dual nature is known as **wave-particle duality.**

Furthermore, scientists have shown that even objects that seem like solid particles, such as electrons or even atoms, can act like waves sometimes. These objects can demonstrate the qualities of interference. For example, two electrons can cancel each other out. Or, the electrons can diffract as they go through an opening. In fact, according to modern physics, *all* objects behave as waves sometimes, and particles sometimes. However, the smaller an object is, the more it acts like a wave. For instance, electrons often act like waves, interfering and diffracting. But, you will never see two baseballs behaving like waves.

Practice 2—Light: Wave or Particle?

Directions: Circle the answer that correctly completes each of the following statements.

1. Experiments show that light travels in bundles of energy known as ____.

    (a) orbits

    (b) photons

    (c) shells

2. A single photon of blue light has ____ a photon of red light.

    (a) more energy than

    (b) less energy than

    (c) the same energy as

3. A brighter light beam consists of ____.

    (a) more photons

    (b) higher energy photons

    (c) more orbits

4. A photon of ____ energy is released when an electron jumps between two orbits that are farther apart.

    (a) higher

    (b) lower

    (c) the same

5. Light is ____.

    (a) a wave

    (b) a particle

    (c) both a wave and a particle

Color

When light from a light bulb is passed through a triangular piece of glass, known as a **prism,** the white light is separated into the colors of the rainbow: red, orange, yellow, green, blue, indigo, and violet. This is because white light is simply the combination of all colors. White itself is not a color. It has no specific frequency or wavelength. It is simply the way our eye perceives a mixture of all colors of light. The prism is capable of "unmixing" the white light. It can separate the white light into its components. When a similar thing happens inside raindrops, a rainbow appears. You will learn how this "unmixing" occurs in Lesson 3.

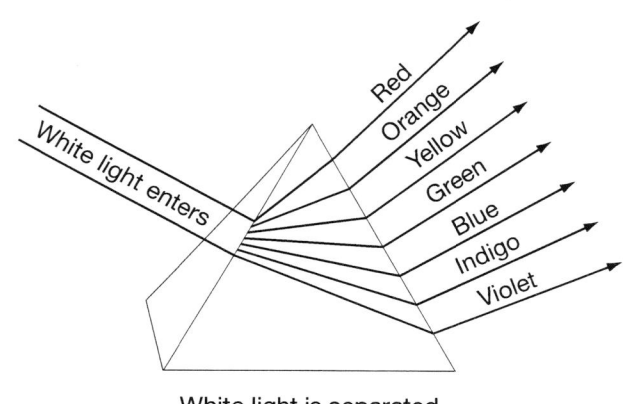

White light is separated into colors as it passes through a prism.

Cones provide us with color vision. Inside your eye, there are three different types of cells, known as cones. One type of cone detects mainly red, another detects mainly green, and another detects mainly blue. When red light comes in, the red cones tell the brain that there is red light. However, if yellow light comes in, both the red *and* green cones are activated. The brain knows that it is yellow light because both the red and green cones are activated.

What if, instead of yellow light, red and green light shine in together? The red and green cones are still both activated. Thus, the brain thinks yellow light is coming in. In fact, if you look at a combination of red and green light, you will see yellow. If you mix red and blue light, you get magenta. If you mix blue and green, you get cyan. If you mix red, green, and blue light all together, you get white light. This is called **additive** mixing of colors.

Because the eye detects red, green, and blue light, these are known as the **additive primary colors.** You can create any other color by using lights of these three colors mixed together. In fact, this is how a color television works. A color television creates red, green, and blue lights on the screen. By mixing these lights together on the screen in different amounts, all the colors that you see on TV can be created.

However, the color of paints, inks, and other pigments is created differently. Red paint appears red because it *absorbs* every color except red. When white light hits it, only red light is reflected into your eyes. That is why the paint appears red.

If you mix red paint and green paint, the red absorbs everything but red, and the green absorbs everything but green. So, in the mixture, all the light is absorbed, and none is reflected. Black is the absence of light. So, the mixture of red and green paint will appear black (or almost black, depending on how pure the paint is). This is known as **subtractive** mixing of colors. Each paint subtracts some of the light that hits it.

The **subtractive primary colors** are magenta, cyan, and yellow. Each of these absorbs only one type of light. When mixing paint, ink, or pigments, any color can be created from these three.

In the 19th century, scientists observed that when certain gases are heated up, they do not glow white. Instead, they emit, or put out, only a few specific colors. This is because, according to Bohr's theory of the atom, electrons in the gases' atoms can jump only between specific orbits. Each jump corresponds to a photon with a certain energy, and thus a specific frequency and color.

What Happens When Red and Green Paints Are Mixed

The set of particular frequencies emitted by a certain type of atom is called its **spectrum.** (The plural of spectrum is spectra.) The spectrum is like a fingerprint for each type of atom. Scientists can use a prism to separate the colors emitted by a glowing gas. This color separation determines the spectrum of the gas. This allows scientists to identify unknown gases. For example, scientists can determine gases that make up the sun or other stars by observing the spectrum of light these stars emit. They can also identify unknown chemicals by burning them and then observing the spectrum of light in the flame.

In Real Life

A neon sign consists of a glass tube filled with neon gas. This gas is heated until it glows. Neon always glows with a characteristic pinkish color. This color is its spectrum. Signs with different colors are made by mixing in other gases, which emit their own spectrum. The mixture of neon's spectrum with other gases' spectra determines the color of the sign.

A laser is a device that creates light of one specific color. Inside a laser, a substance is caused to glow. This substance then emits a characteristic spectrum. One frequency increases until it is the only color emitted.

Because only one color is produced, the light beams all have the same wavelength. Furthermore, the method used to amplify this color ensures that all of the beams are "in step" with each other, or in phase. This is known as **coherent light.** On the other hand, a lightbulb puts out incoherent light. **Incoherent light** consists of many different wavelengths, all out of phase with each other. The coherence of laser light is important in making three-dimensional photographs called holograms. But, lasers also have many other applications

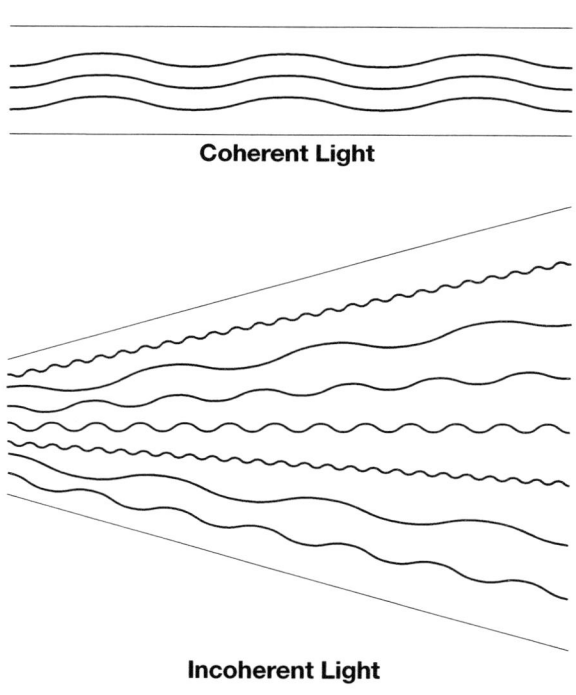

Coherent Light

Incoherent Light

In addition to being in phase, all of the light from a laser is traveling in the same direction. The light from a flashlight or lightbulb spreads out as it travels. On the other hand, light from a laser continues to travel in a straight line. This allows all of its energy to be concentrated in one spot. For example, light from a laser has been reflected off mirrors left by astronauts on the moon. The light from a normal lamp would spread out too much for this to be possible. This is also useful in laser surgery. A laser beam is used to cut by focusing all of its energy on one tiny spot. Lasers can also be used in communication because the beam can travel over extremely long distances without spreading out. This is done using optical fibers. **Optical fibers** are thin strands made of glass or plastic that can carry laser light, much like a pipe carries water, or a wire carries electricity.

Practice 3—Color

Directions: Circle the answer that correctly completes each of the following statements.

1. When light passes through a ____, it is separated into its component colors.

   (a) prism   (b) optical fiber   (c) laser

2. ____ is the combination of all colors of light.

   (a) White   (b) Black   (c) Red

3. Red, green, and blue are the ____ primary colors.

   (a) additive   (b) subtractive   (c) divisive

4. Blue paint appears blue because it ____ blue light.

   (a) absorbs   (b) transmits   (c) reflects

5. Mixing paints is an example of ____ mixing of colors.

   (a) additive   (b) subtractive   (c) divisive

Directions: Decide whether each statement that follows is true (**T**) or false (**F**).
Write the correct letter in each blank.

____ 6. Your eye contains one type of cone for every color in the rainbow.

____ 7. When a gas is heated, it puts out a unique set of colors known as its spectrum.

____ 8. A laser produces incoherent light, which spreads out quickly.

■ LESSON MASTERY TEST

Directions: Circle the answer that correctly completes each of the following statements.

1. Light is a ____.

   (a) wave   (b) particle   (c) wave and a particle

2. A bundle of electromagnetic energy that acts like a particle is called a ____.

   (a) vibration   (b) photon   (c) orbit

3. Electromagnetic waves are created by the vibration of ____.

   (a) atoms   (b) speakers   (c) electrons

4. The more energy a photon has, the ____ its frequency.

   (a) higher   (b) lower

5. A brighter light beam consists of ____.

   (a) more photons

   (b) higher energy photons

   (c) more orbits

Directions: Decide whether each statement that follows is true (**T**) or false (**F**). Write the correct letter in each blank.

____ 6. Electromagnetic waves cannot travel through empty space because they require air as a medium.

____ 7. Red, green, and blue are the additive primary colors.

____ 8. Light travels more slowly through glass than through air.

____ 9. The color of a T-shirt is determined by the colors of light it reflects.

____ 10. Many kinds of gas produce the same spectrum.

____ 11. The light waves produced by a laser are all in phase, with the same frequency.

■ Lesson 3—The Behavior of Light

Goal: To understand how light travels and interacts with objects

Rays and Reflection

In the last lesson, you learned that the nature of light is very complex. Light can behave like a particle or a wave. However, scientists often take a simple view of light in order to understand how it travels and interacts with everyday objects. **Optics** is the study of how visible light interacts with everyday objects.

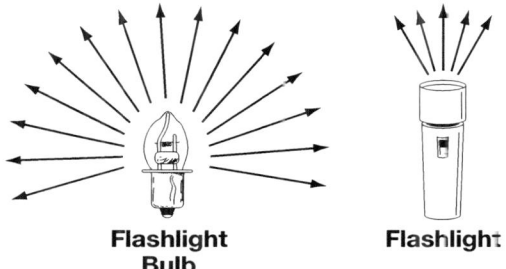

Flashlight Bulb　　**Flashlight**

When you turn on a flashlight, the beam of light travels out in straight lines from the bulb. We call these lines rays. A **ray** is simply the path of the light. You don't need to worry about whether a particle or wave is traveling along this path.

In the diagram below, you can see light rays hitting three different objects. Ray A hits a piece of glass, and passes through it. This is called **transmission.** Ray B hits a piece of black velvet, and is stopped. This is known as **absorption.** Finally, ray C hits a piece of shiny metal and bounces off. This is **reflection.**

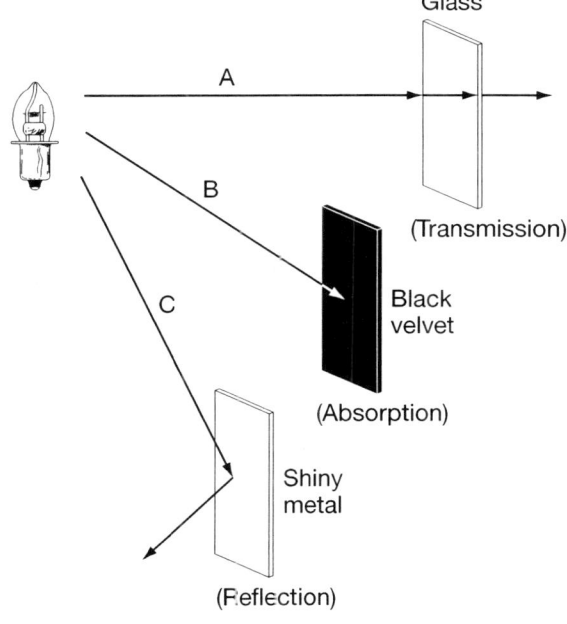

When light reflects off a surface, the angle at which it strikes the surface is known as the **angle of incidence.** The angle at which light leaves a surface is known as the **angle of reflection.** These angles are both measured against the line that is perpendicular to the surface. This line is called the **normal.** These lines and angles are shown in the diagram at the right.

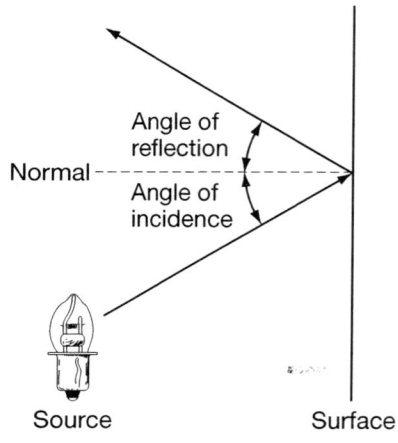

Whenever light reflects off a surface, the angle of incidence is always *equal to* the angle of reflection. This is known as the **law of reflection.**

When you look in a mirror, you see a very clear image of yourself. This is known as **specular reflection.** This occurs in any extremely smooth surface. You can even see a reflection on a smooth table or a window.

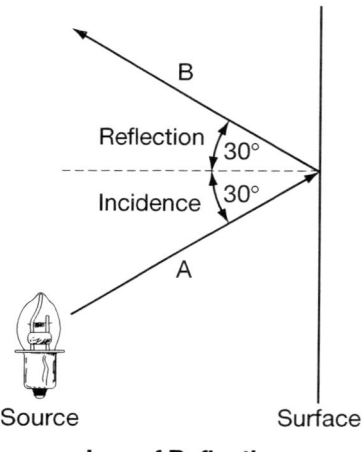

Law of Reflection

You can see an object only if light reflects off it into your eyes. So, why can't you see specular reflections in all objects, as you do in a mirror? If a surface is rough, the light hits the surface at many different angles of incidence. It then reflects in every direction. This is known as **diffuse reflection.** Even objects that seem smooth, like a piece of paper, are actually extremely rough. Light rays are tiny compared with the hills and valleys on a piece of paper.

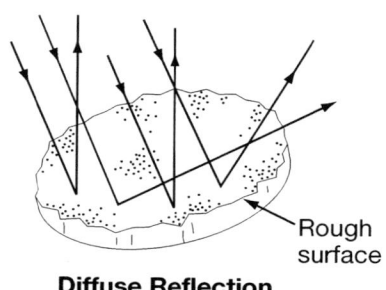

Diffuse Reflection

Practice 1—Rays and Reflection

Directions: Circle the answer that correctly completes each of the following statements.

1. When light passes through an object, we say it is ____.

   (a) reflected   (b) absorbed   (c) transmitted

2. A line representing the path light follows is called a ____.

   (a) ray   (b) wave   (c) reflection

3. When light reflects off a surface, the angle of incidence is always ____ the angle of reflection.

   (a) less than   (b) equal to   (c) greater than

4. When light reflects off a smooth surface like a mirror, we call it ____ reflection.

   (a) specular   (b) diffuse   (c) normal

5. When light reflects off a rough surface, it is called ____ reflection.

   (a) specular   (b) diffuse   (c) normal

Refraction and Lenses

When light passes through a transparent material, such as glass, it is slowed down. This is because the molecules in the glass impede, or slow down, the light's motion. When the light hits a molecule in the glass, it is actually absorbed, and then re-emitted. This causes the light to travel more slowly than if it were moving freely. Remember, light only travels at 75% of its full speed in water. And, it only travels at 65% of its full speed in glass. In general, the more dense a substance is, the more it will slow down light.

This change in speed causes light to bend when it hits a transparent substance. To understand why, imagine that you are driving in a car on a road, and the side of the road is gravel. If your right tires hit the gravel, the right side of the car will suddenly slow down. This makes the car turn toward the gravel.

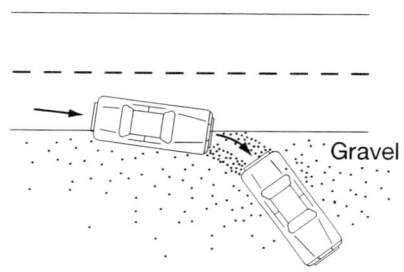

Similarly, when a light wave hits the surface of a piece of glass, one part of the wave is slowed down first. This makes the wave turn into the surface. On the other hand, when light goes from a dense (slow) substance into a less dense (faster) substance, such as from glass to air, it bends outward. If the wave hits the surface straight on, there is no bending. This is because no part of the wave hits first. The bending of light as it passes between different transparent substances is called **refraction.**

The diagram at the right shows light passing from air into glass. The dashed lines perpendicular to the surface are the normal lines. The **law of refraction** states that:

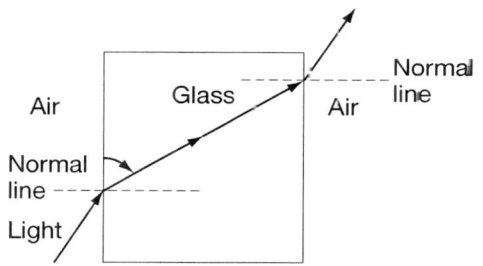

- Light passing into a slower medium will be bent *toward* the normal.

- Light passing into a faster medium will be bent *away from* the normal.

The more the light changes speed, the more it will bend when crossing the surface.

You can see the effect of refraction if you put a spoon into a glass of water. It will appear that the spoon bends at the surface of the water. In fact, it is simply the light moving from water into air that bends, not the spoon.

Refraction is also responsible for the mirages of water you may sometimes see on hot roads. On warm days, the air right above the road becomes very hot. Light travels faster through hot air, so it bends. When you look at the road, you are actually seeing a reflection of the sky. This reflection is bent by the hot air above the road.

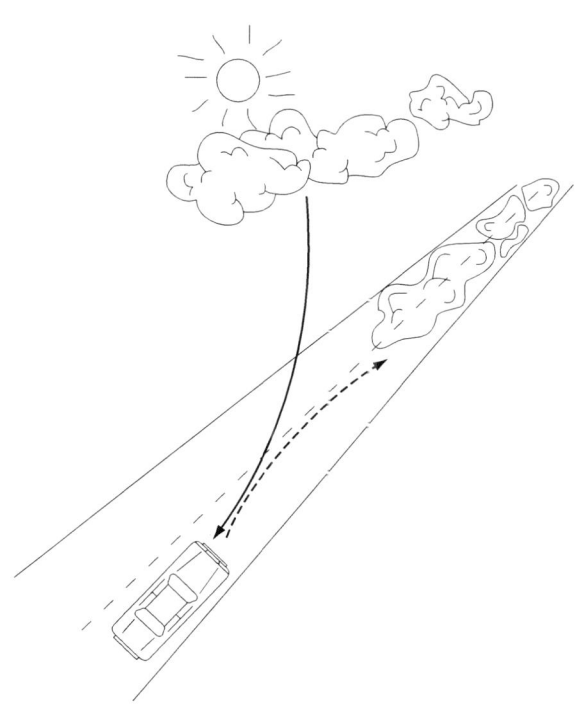

Lenses use refraction to bend light. This bent light creates images that are larger or smaller than the original object. There are two main types of lenses: converging and diverging. In a **converging lens,** parallel beams of light are brought together at one point, known as the **focal point.** The distance between the lens and the focal point is known as the **focal length.** The shorter the focal length, the more the light is bent. This is what makes the lens more powerful.

On the other hand, a **diverging lens** spreads out the light that passes through it. Its focal point is where the beams appear to be diverging from.

Your eye contains a converging lens that focuses incoming light on the back of the eye. This surface at the back of the eye is known as the **retina.** In people who cannot see clearly, the lens focuses the light either in front of or behind the retina instead of on the retina. That is why they cannot see clearly. Glasses correct this by causing the light to be bent more or less *before* it reaches the eye.

Recall that a prism separates white light into different colors using refraction. Different colors of light actually travel at slightly different speeds in glass. This means that when light passes from the air into the prism, different colors of light are bent by different amounts. When they leave the prism, they come out at different angles. The different refraction for different colors is known as **dispersion.**

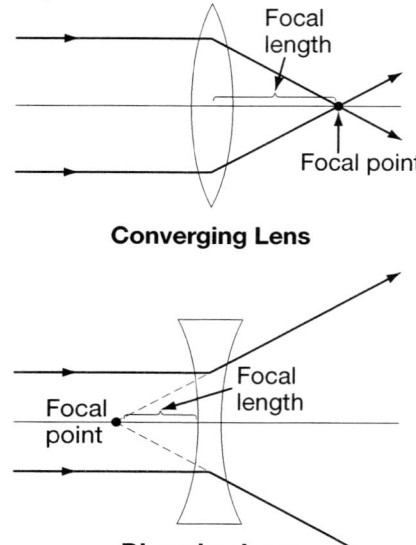

Converging Lens

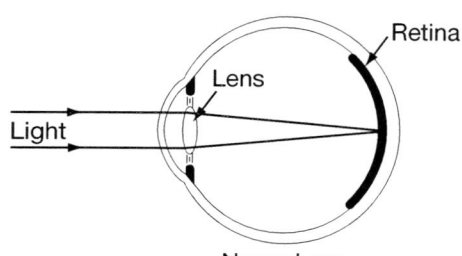

Diverging Lens

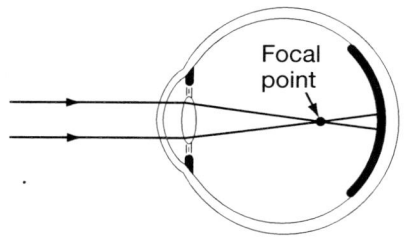

Normal eye

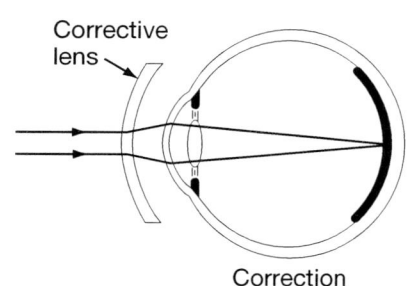

Cannot see clearly

Correction

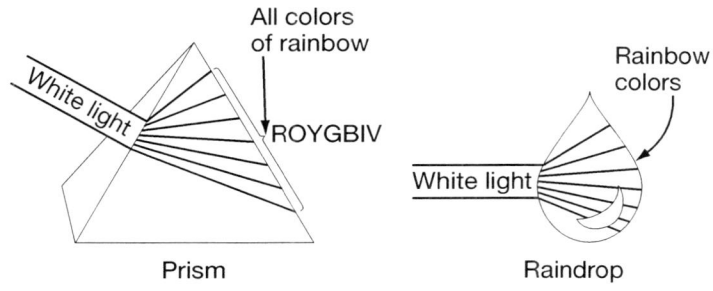

All colors
of rainbow

ROYGBIV

Rainbow
colors

White light

White light

Prism

Raindrop

Dispersion

Dispersion is also responsible for a rainbow. When white light hits a raindrop, it is bent once as it enters the drop. It then reflects off the back of the raindrop, and is bent again as it comes out. The different colors are bent by different amounts. Thus, the white light of the sun separates into the colors of the rainbow. The colors of the rainbow are red, orange, yellow, green, blue, indigo, violet (ROYGBIV).

Practice 2—Refraction and Lenses

Directions: Circle the answer that correctly completes each of the following statements.

1. Light _____ when entering a denser medium.

    (a) speeds up

    (b) slows down

    (c) stops

2. The bending of light when entering a different medium is known as _____.

    (a) refraction

    (b) reflection

    (c) diffraction

3. When lights enters a denser medium, it _____.

    (a) bends away from the normal

    (b) bends toward the normal

    (c) doesn't bend at all

(continued on next page)

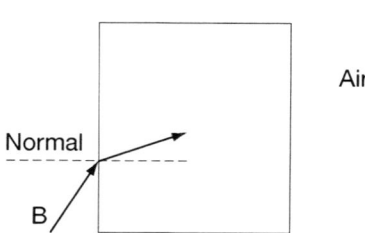

4. The diagram above shows light entering a ____ medium.

   (a) denser   (b) less dense

5. Light travels ____ in hot air, which results in mirages on hot days.

   (a) faster   (b) slower   (c) the same speed

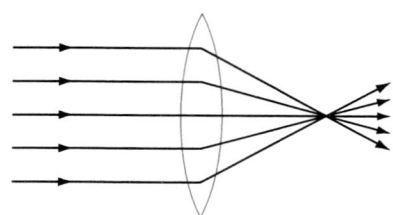

6. The diagram above shows a ____ lens.

   (a) diverging   (b) converging

7. The longer the focal length of a lens, the ____ it bends light.

   (a) more   (b) less

8. Different colors of light travel at ____ speeds in glass.

   (a) different   (b) the same

9. When different colors of light refract by different amounts, as in a prism or rainbow, it is called ____.

   (a) diffraction   (b) dispersion   (c) reflection

Interference and Polarization

Remember from Lesson 1 that two waves which are in step (in phase) with each other add up to make a bigger wave. But, waves that are out of step (out of phase) cancel each other out. This effect, interference, can be seen with light. Remember, light acts like a wave.

Sometimes you can see a rainbow effect in soap bubbles, or in a puddle with an oily surface. Both of these are the result of interference. When light hits a thin film, such as a soap bubble, or oil on top of water, it reflects in two different ways. Some of the light is reflected off the top of the film. And, some is reflected off the bottom of the film.

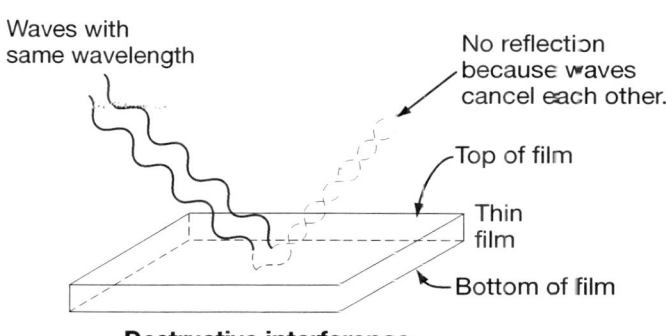

Destructive interference

When the two reflected waves overlap, there will be interference. Depending on the thickness of the film, the reflections of the waves will either be in phase or out of phase. If they are in phase, that color will appear bright. If they are out of phase, that color will not appear.

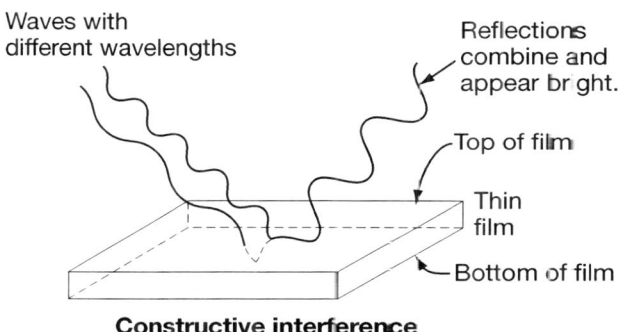

Constructive interference

Because different colors have different wavelengths, some colors will interfere destructively, and some will interfere constructively. Only the colors that undergo constructive interference will be seen. So, they determine the color of the reflection. However, because the film has slightly different thicknesses in different places, different colors will appear in different areas of the film. This is what produces the rainbow. The production of color by interference is known as **iridescence.**

Another wave effect in light is polarization. When you create waves in a rope, the waves can be horizontal, vertical, or somewhere in between. This is known as the **polarization** of the wave.

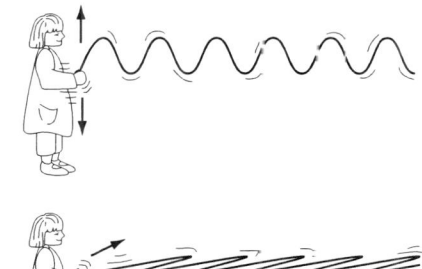

Similarly, light waves have a polarization. However, light from a flashlight or the sun contains light waves polarized in all different directions. We call this **unpolarized** light.

Some substances only allow light with a certain polarization to pass through. Think of shaking the rope through an open slot in a picket fence. Only vertical waves can pass through the slot in the fence. Other waves are stopped when they collide with the pickets in the fence. Polarizing filters, such as polarized sunglasses, only allow light waves with the proper polarization to pass through.

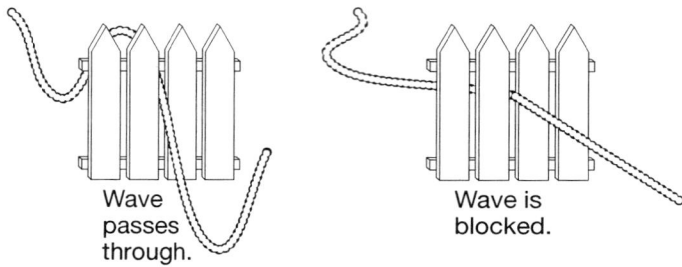

Wave passes through.

Wave is blocked.

Think About It

What would happen if you looked through two pairs of polarized sunglasses at the same time? What if you turned the sunglasses at right angles to each other?

When light reflects off a horizontal surface, such as a lake or snow, it also becomes partially polarized. Polarized sunglasses are very useful for blocking glare from such surfaces. Light reflecting off this kind of surface is polarized horizontally. Vertically polarized sunglasses will block the reflected light. But, they will let other light through. Similarly, sunlight reflecting off snow (or even other cars when you are driving) is generally horizontally polarized. Vertically polarized sunglasses will also block this glare.

Practice 3—Interference and Polarization

Directions: Decide whether each statement that follows is true (**T**) or false (**F**). Write the correct letter in each blank.

_____ 1. Light waves can never undergo destructive interference.

_____ 2. When two waves that are out of phase overlap, they add up to produce a brighter wave.

_____ 3. In a thin film, colors can destructively interfere or constructively interfere.

_____ 4. The only colors you see reflected from a bubble are those that interfere constructively.

_____ 5. A flashlight creates light that is polarized in only one direction.

_____ 6. Similar polarizing filters that are placed at right angles to each other will completely block light from coming through.

_____ 7. Sunglasses that are polarized vertically will block light that is reflected off horizontal surfaces.

Blue Skies, Red Sunsets

When light hits small particles, such as gas molecules in the atmosphere, the light can bounce off the particles. Then the light is re-emitted in all different directions. This is known as **scattering**.

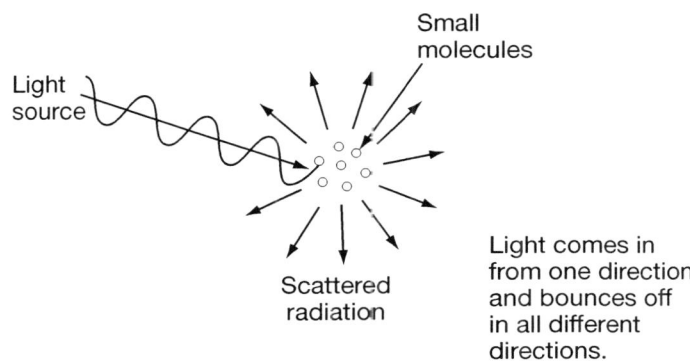

Light source

Small molecules

Scattered radiation

Light comes in from one direction and bounces off in all different directions.

When you look up at the sun, you are seeing light that has passed straight through the atmosphere. However, there is nothing creating light in other parts of the sky. You only see light that is scattering when it bounces off air molecules. Not all colors of light scatter the same amount. Because the short wavelength of blue light is closest to the size of air molecules, it is scattered the most. So, most of the light reflecting off other parts of the sky is blue, making the sky appear blue.

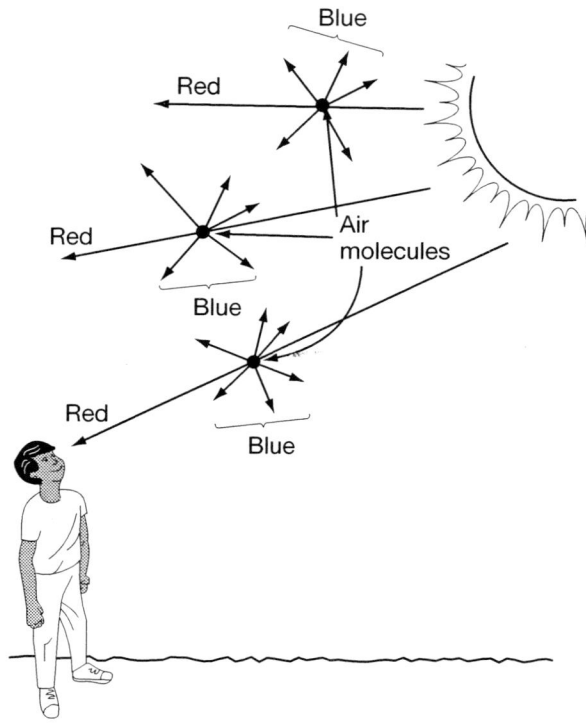

Red light comes straight to person. Blue light scatters. Since blue light scatters and red does not, the sky appears blue.

The sunset appears red for a similar reason. When the sun is directly overhead, its light is taking the shortest path through the atmosphere. The sun appears white because almost all of the light can pass through the atmosphere.

Right before sunset, however, the light must pass through much more of the atmosphere to reach your eyes. As the light passes through the atmosphere, some of it scatters off the air molecules. Because red has the longest wavelength, it scatters the least. So, most of the light that reaches your eyes is red. Thus, the sun appears red at sunset.

Dust or other pollution in the air can make the sunset particularly colorful. This is because extra particles in the air scatter more of the light. Only the very lowest frequency red light can get through, making the sun appear a fiery red at sunset. Forest fires and volcanoes, which both send a lot of particles into the atmosphere, often result in spectacular sunsets.

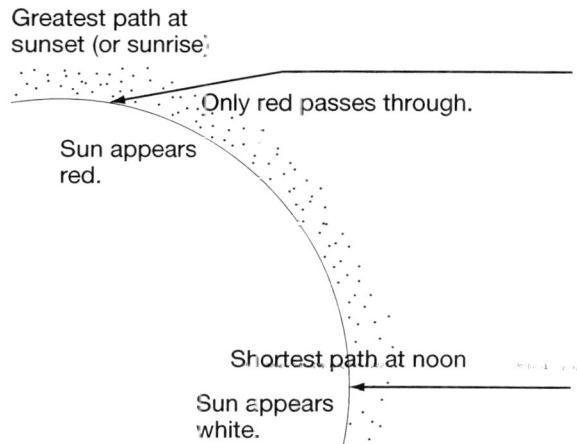

Greatest path at
sunset (or sunrise)

Only red passes through.

Sun appears
red.

Shortest path at noon

Sun appears
white.

When you are looking at the sky during the daytime, it appears blue because you are seeing the blue light that is scattered. When you look at the sun when it is setting, you are seeing only the light that does not get scattered, which is red.

Think About It

You can make your own blue sky and red sunset. Put a few drops of milk in a glass of water. Then, turn off the lights, and shine a flashlight into the glass. If you look from the side perpendicular to the light, the glass will appear blue. If you look from the side that is directly opposite the flashlight, the glass will appear red. Why does this work?

Practice 4—Blue Skies, Red Sunsets

Directions: Circle the answer that correctly completes each of the following statements.

1. When light bounces off small particles and is sent in many different directions, it is known as ____.

   (a) reflection

   (b) dispersion

   (c) scattering

2. Blue light scatters ____ red light.

   (a) more than

   (b) less than

   (c) the same amount as

3. When you see the blue sky, you are seeing the blue light that is ____ by air molecules.

   (a) created

   (b) absorbed

   (c) scattered

4. When the sun looks red at sunset, you are seeing the red light that ____ by air molecules.

   (a) is scattered

   (b) is not scattered

   (c) is absorbed

5. Dust and other particles in the atmosphere make the sunset appear ____.

   (a) more red

   (b) darker

   (c) earlier

■ LESSON MASTERY TEST

Directions: Circle the answer that correctly completes each of the following statements.

1. When a ray of light bounces off an object, we say it is _____.

    (a) reflected   (b) absorbed   (c) transmitted

2. When light hits a surface and reflects in many directions, we call it _____ reflection.

    (a) specular   (b) diffuse   (c) normal

3. The line perpendicular to a surface is called the _____.

    (a) right line   (b) normal line   (c) angle of incidence

4. Light slows down when entering _____.

    (a) a more dense medium   (b) a less dense medium

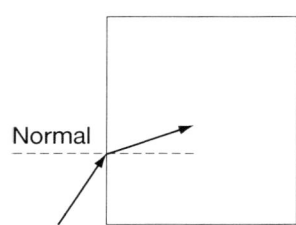

5. The diagram above shows light entering a _____ medium.

    (a) denser   (b) less dense

6. In a converging lens, parallel beams of light are brought together at the _____.

    (a) focal length   (b) focal point   (c) converging point

(continued on next page)

7. In a thin film, the only colors that can be seen are those that interfere ____.

   (a) constructively (b) destructively (c) diffusely

8. When light bounces off small particles and is sent in many different directions, it is known as ____.

   (a) reflection (b) dispersion (c) scattering

Directions: Decide whether each statement that follows is true (**T**) or false (**F**). Write the correct letter in each blank.

_____ 9. At sunset, the sun appears red because red light is scattered the most.

_____ 10. The light produced by a flashlight is polarized in one direction only.

_____ 11. When entering a denser medium, light bends toward the normal line.

_____ 12. When two waves that are out of phase overlap, they add up to produce a brighter wave.

_____ 13. Sunglasses that are polarized vertically will block light that is reflected off horizontal surfaces.

_____ 14. Light can bend when passing from cold air into hot air.

_____ 15. When light reflects off a surface, the angle of incidence is always equal to the angle of reflection.

PART 4

ELECTRICITY, MAGNETISM, AND BEYOND

Table of Contents

■ Lesson 1—The Basis of Electricity and Magnetism

Goals: To understand the subatomic basis of electricity and magnetism; to understand the interaction between electricity and magnetism

Charges and Electric Current

In this lesson, you will learn about electricity and magnetism. Both are the result of electric charges. There are two types of electric charge: positive and negative. Charge is not something that can be seen or touched. Charge is a property of particles. Two particles with the same charge repel, or push away from, each other. Two particles with opposite charges attract each other. This is known as the **electric force.**

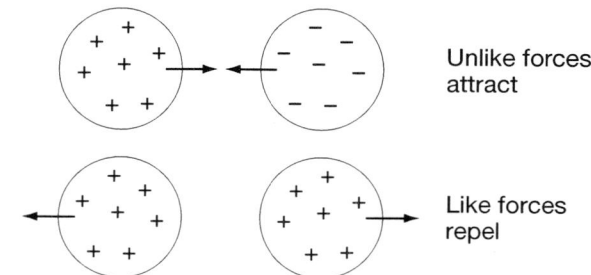

Electric Force

In the atom, there are two types of particles that have charge. Protons have positive charge. Electrons have negative charge. The force of attraction between the negative electrons and positive protons keeps the atom together. As long as the atom has the same number of electrons and protons, it will have no overall charge. This is because the negative electrons cancel the positive protons. An object with no overall electric charge is called **neutral.** For example, your body is neutral because it has the same number of electrons and protons.

Electric charge is measured in **coulombs (C).** One coulomb is a very large amount of charge. For example, the typical charge when you rub a balloon against your shirt is less than a millionth of a coulomb, and the charge on one electron is much smaller than that, around 2×10^{-19} C.

Because the electrons are on the outside of the atom, sometimes they can be removed from the atom. They can also jump from one atom to another. This is especially true in metals. In metals, the electrons can easily drift from one atom to another.

When an atom loses an electron, the atom becomes positively charged. This is because it now has more positive protons than negative electrons. On the other hand, if an atom gains an electron, it becomes negatively charged. For example, if you run a comb through your hair very quickly, you can rub some of the electrons off of your hair and onto the comb. This makes your hair positively charged and the comb negatively charged. Your hair will then stick to the comb because of the electric force. **Static electricity** happens when charges are transferred due to rubbing or other motion.

 How many everyday examples of static electricity can you think of?

On the other hand, sometimes electrical energy is used to push the electrons. For example, when you connect a lightbulb to a battery, the battery is supplying a push to the electrons in the lightbulb and the wires. This forces electrons to move through the wires and the lightbulb. When you plug a television into the outlet, the power from the electric plant pushes the electrons through the television.

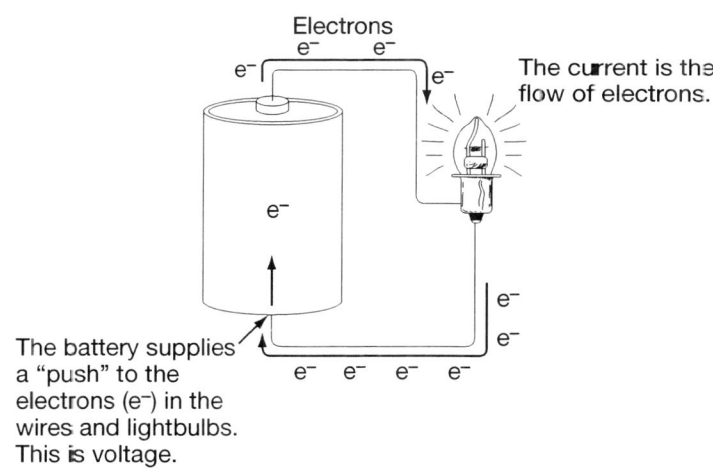

Electrons

The current is the flow of electrons.

The battery supplies a "push" to the electrons (e⁻) in the wires and lightbulbs. This is voltage.

The flow of electrons is known as a **current.** The more electrons that move, and the faster they move, the greater the current. Current is measured in **amperes (A),** often abbreviated as **amps.** An electric shock of half an amp can kill a person. Typical current in a flashlight is around one hundredth of an amp. Because the amp is so large, current is sometimes measured in **milli-amps (mA).** One milli-amp is equal to one thousandth of an amp.

The push given to electrons is known as an **electric potential,** or **voltage.** It is measured in **Volts (V).** The voltage written on a battery, 1.5 V, 9 V, or 12 V, indicates how much electrical energy the battery gives to each electron.

It is important to remember that a battery, or any other voltage source, is not actually supplying electric charge. It is only giving a push to the electrons that are already in the wires, and to the electrons in whatever appliance is being used. All conducting substances contain loose electrons. But, these electrons will only move when energy is supplied to them by the voltage source.

In addition to a voltage source, electrons must have a complete path to travel. For example, in a flashlight, the electrons travel from the battery to the lightbulb, and back into the battery. A complete path such as this is called a **closed circuit.** If one of the wires from the lightbulb to the battery were missing, the path would not be complete, and electricity would not flow. This is known as an **open circuit.**

If there is a voltage source and a closed circuit, electrons will flow. As the electrons move through an electrical appliance, such as a television, a stereo, or an electric mixer, their electrical energy is converted into other forms. These forms are the picture on the television, the sound of the stereo, or the motion of the electric mixer. This is why electricity is so useful.

Practice 1—Charges and Electric Current

Directions: Circle the answer that correctly completes each of the following statements.

1. Electric current is measured in _____.

   (a) amps   (b) volts   (c) coulombs

2. The push of energy given to electrons in a circuit is known as ____.

   (a) current   (b) voltage   (c) charge

3. A complete path for electricity is known as a(n) ____.

   (a) open circuit   (b) closed circuit (c) neutral circuit

4. A wire already contains charges, and a battery simply provides the ____ to make the electrons move.

   (a) push   (b) pull   (c) static

Resistance and Ohm's Law

Objects that let electricity flow through them very easily are called **conductors.** In conductors, the electrons are not held tightly to the atom. There is very little friction as the electrons move through the material. Metals are very good conductors of electricity. On the other hand, **insulators,** such as plastic and wood, do not allow electrons to move easily. It takes a large push, or high voltage, to move electrons through an insulator.

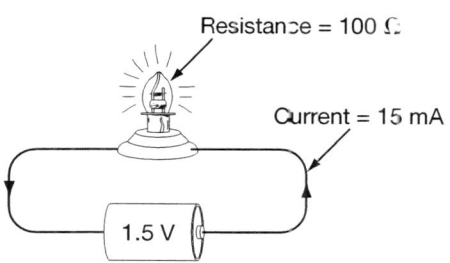

Resistance = 100 Ω
Current = 15 mA
1.5 V

Resistance is the measurement of how difficult it is for electrons to move through an object. The unit for resistance is the **Ohm,** abbreviated with the Greek letter Ω **(omega).** Metals and other insulators have a very low resistance, while insulators have a very high resistance. Similarly, a long piece of wire has more resistance than a short piece of wire. This is because there is a longer distance through which the electrons must be pushed.

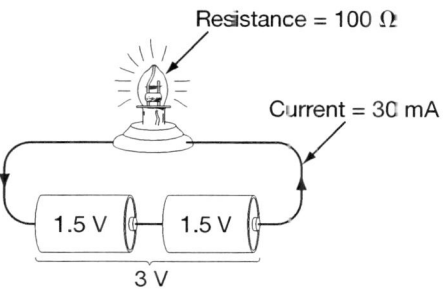

Resistance = 100 Ω
Current = 30 mA
1.5 V 1.5 V
3 V

The voltage and resistance together determine how much current will flow in a circuit. Increasing the voltage will cause a greater current to flow. On the other hand, a large resistance, such as wire made from a poor conductor, will cause less current to flow.

Same resistance with greater voltage means more current.

You can predict the flow of electricity (amperes) in a circuit by using an equation known as **Ohm's Law.**

$$V = I \times R$$

Here V stands for voltage, I stands for current, and R stands for resistance.

Example 1: If a lightbulb has a resistance of 150 Ω, and you connect it to a 1.5 V battery, how much current will flow?

$$V = I \times R$$

$$1.5 \text{ V} = I \times 150 \text{ Ω}$$

Divide both sides by 150 Ω.

$$I = .01 \text{ A}$$

So, one-hundredth of an amp, or 10 mA, will flow.

Example 2: If a lightbulb has a resistance of 200 Ω, and you want .03 A of electricity to flow through it, what voltage battery should you use?

$$V = I \times R$$

$$V = .03\ \text{A} \times 200\ \Omega$$

$$V = 6\ \text{V}$$

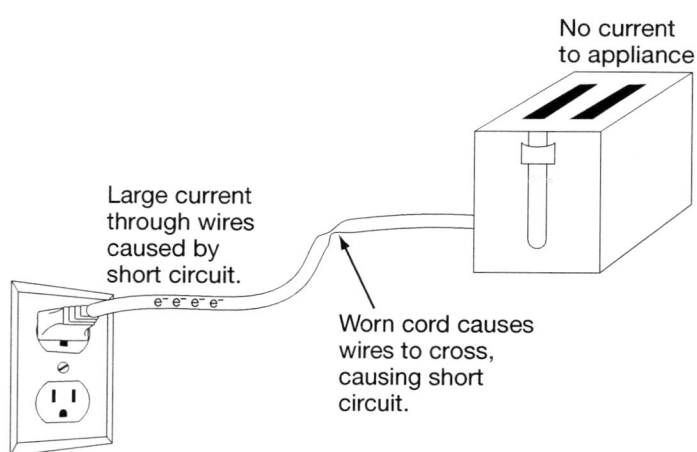

No current to appliance

Large current through wires caused by short circuit.

e⁻ e⁻ e⁻ e⁻

Worn cord causes wires to cross, causing short circuit.

You should use a 6-volt battery.

The part of the circuit that uses electricity, such as a lightbulb or television, is known as the **load.** Usually, the load has the most resistance in a circuit. And, the wires leading to the load have almost no resistance. If a circuit is connected without a load, there is very little resistance. So, a large amount of current can flow. This is known as a **short circuit.** A short circuit can be very dangerous.

In Real Life

In your house, the cord for an old appliance sometimes wears out and the wires can become crossed. Then, electricity can flow through the wires without going through the appliance. Because there is so little resistance, a large current flows. The heat created by the current can be great enough to start a fire. Therefore, houses have **circuit breakers,** or fuses. These fuses detect when too much current is flowing, and cut off the voltage supply.

Recently, substances have been discovered that have absolutely no resistance to the flow of electricity. It takes virtually no energy to create a current in these substances, which are known as **superconductors.** Such substances have many potential applications in electronics and machinery. However, at the present time, these substances only work as superconductors at extremely low temperatures. This means they must be cooled to below −200°C using liquid nitrogen. This makes superconductors very impractical. But, scientists are currently working to develop substances that could be superconductors at room temperature.

Other substances have properties that fall between a conductor and an insulator. These are known as **semiconductors.** Semiconductors are very important in computers and other electronic devices because they can be made very small and don't require a lot of power. When they are used in circuits, semiconductors conduct electricity only under specific circumstances. The transistors in a transistor radio are made from semiconductive material.

Practice 2—Resistance and Ohm's Law

Directions: Circle the answer that correctly completes each of the following statements.

1. A material that lets electricity flow through it easily is known as a(n) ____.

    (a) insulator   (b) conductor   (c) circuit breaker

2. An object that lets electricity flow easily has ____ resistance.

    (a) high   (b) low   (c) even

3. Applying a greater voltage to a circuit allows ____ current to flow.

    (a) more   (b) less   (c) the same

4. Adding more resistance in a circuit allows ____ current to flow.

    (a) more   (b) less   (c) the same

5. ____ of electrical energy is needed to cause .5 A of current to flow in a circuit with 20 Ω of resistance. (Remember Ohm's Law: $V = I \times R$)

    (a) 5 V   (b) 10 V   (c) 20 V

6. If a 9 V battery is connected to a lightbulb with 4.5 Ω of resistance, ____ of current will flow. (Again, remember Ohm's Law.)

    (a) 1 A   (b) 2 A   (c) 9 A

7. A circuit which allows current to flow, but not flow through the load, is known as a(n) ____.

    (a) open circuit   (b) closed circuit   (c) short circuit

8. A(n) ____ protects your house in case too much current flows through the wires.

    (a) circuit breaker   (b) insulator   (c) conductor

9. A substance which behaves as a conductor and an insulator is known as a ____.

    (a) superconductor   (b) semiconductor   (c) resistance

Magnetism

If you have ever held two magnets near each other, you've noticed that if they are held one way, they attract, while if they are held the other way, they repel. This is because each magnet has two sides, known as **poles.** One side is known as the north pole, and the other is the south pole. Magnetic poles act much like electric charges. This means that similar poles repel each other, and opposite poles attract each other.

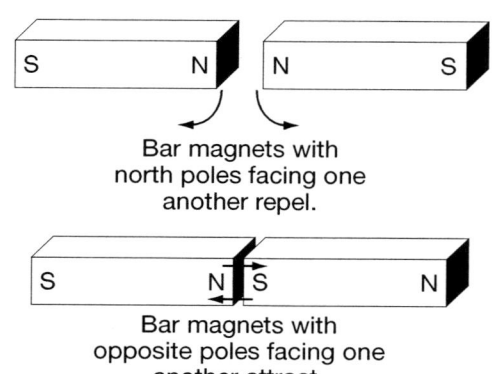

Bar magnets with north poles facing one another repel.

Bar magnets with opposite poles facing one another attract.

However, magnetic poles are different than electric charges. This is because electric charges can be separated. For instance, an electron has only a negative charge, and a proton has only a positive charge. However, all magnets have both a north pole and a south pole. You cannot separate them. If you take a bar magnet and break it in half, you will not get a separate piece of north pole and a separate piece of south pole. Instead, each half will have both a north pole and a south pole. You can continue breaking the magnet, but each time you do this, you will get another complete magnet.

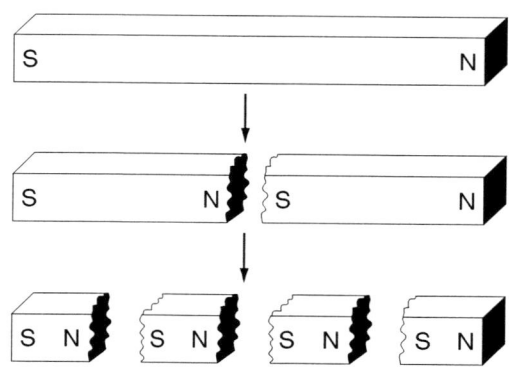

A bar magnet with a north and a south pole will continue to maintain north and south poles no matter how many times it is broken.

The needle of a compass is a tiny magnet. It always points toward north. This is because Earth itself is a very large magnet. What we call the geographic North Pole is actually the south pole of Earth's magnetic field. The magnetic north pole of the compass needle is attracted to Earth's South Pole. The other end is repelled. This causes the needle to line up with Earth's magnetic field. Similarly, the compass needle will line up with any other magnet's north and south poles. If you hold a bar magnet near a compass, the compass will no longer point toward the North Pole. Instead, it will line up to the bar magnet.

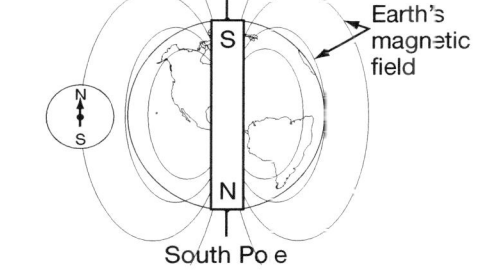

We often represent the magnetic force around a magnet by its **magnetic field.** The magnetic field is a set of lines showing the direction of the magnetic force at that point. A compass needle will line up to the magnetic field. Furthermore, the closer together the magnetic field lines are, the stronger the magnetic force. The diagram below shows the magnetic field of a bar magnet. Notice how the lines are closer together near the poles, where the force is stronger.

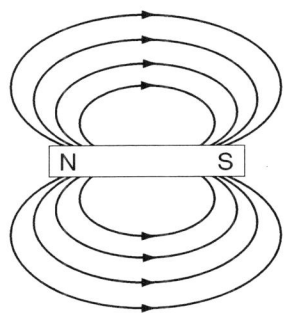

A magnet can cause nearby metals to become magnetized as well. If you stick a magnet on a nail, after a few minutes the nail will become magnetized. It will be able to pick up small paper clips, and it will attract a compass needle. However, after a few hours, it will lose its magnetism. On the other hand, substances that do not quickly lose their magnetism are known as **permanent magnets.** The magnets that you buy in the store are permanent magnets.

Practice 3—Magnetism

Directions: Decide whether each statement that follows is true (**T**) or false (**F**).
Write the correct letter in each blank.

_____ 1. The north poles of two magnets will attract each other.

_____ 2. A magnet has either a north pole or a south pole, but not both.

_____ 3. If you break a magnet in half, each piece will have both a north pole and
a south pole.

_____ 4. A compass needle always points to the north pole of Earth's magnetic field.

_____ 5. The magnetic field represents the pattern of magnetic force around an object.

_____ 6. Once an object is magnetized, it will remain magnetized forever.

_____ 7. It is possible to magnetize some materials by placing them near another magnet.

Electric Currents and Magnetism

In 1820, Danish scientist Hans Christian Orsted was demonstrating electricity and magnetism to a university class. He noticed that every time he connected a wire to a battery near a compass, the compass needle would move. Orsted realized that the current flowing in the wire was creating a magnetic field.

A straight line of wire that is carrying current creates a circular magnetic field. When this wire is formed into one or more loops, a different magnetic field is produced. It turns out that the magnetic field produced by the looped wire is the same as the magnetic field of a bar magnet.

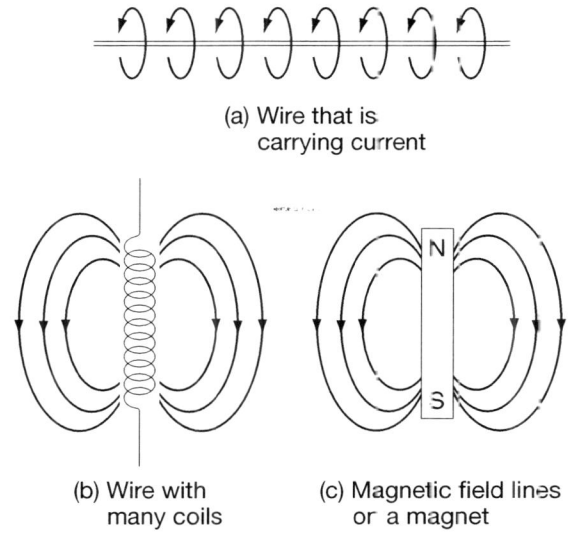

(a) Wire that is carrying current

(b) Wire with many coils

(c) Magnetic field lines or a magnet

The more current there is in the wire, the stronger the magnetic field that is produced. And, the more times you loop the wire around, the stronger the magnetic field. A magnet created by many coils of wire is known as an **electromagnet.**

In Real Life

You can make your own electromagnet by wrapping a wire around a nail many times. The metal in the nail concentrates the magnetic field. If you connect the wire to a battery, you can use the nail to pick up small pieces of metal. If you use a stronger battery to produce more current, the electromagnet will be more powerful. Likewise, if you wrap the wire around more times, the magnet will be stronger.

In fact, all magnets are a result of electric currents. However, in magnetic substances, such as the materials bar magnets are made from, the electric currents occur in single atoms.

Remember that in the atom, the electrons are circling around the nucleus. Each circling electron is like a tiny loop of electric current. Thus, each electron produces a magnetic field. This is like a tiny electromagnet. Thus, each atom can be a magnet.

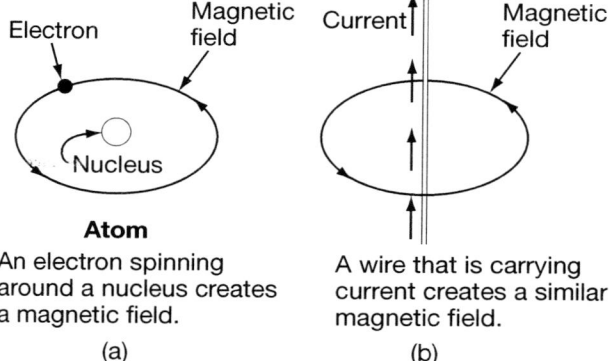

An electron spinning around a nucleus creates a magnetic field.

(a)

A wire that is carrying current creates a similar magnetic field.

(b)

In most materials, however, the magnetic field is canceled out in one of two ways. All atoms, except for hydrogen, contain more than one electron circling the nucleus. Each circling electron produces its own magnetic field. However, electrons circling in opposite directions produce opposite magnetic fields. Therefore, they cancel each other. Many materials are not magnetic because the magnetic fields produced by the electrons in the atom cancel each other out.

In other materials, the magnetic fields created by neighboring atoms cancel out. For example, in iron, each atom has its own magnetic field. However, the orientation of each atom's magnetic field is different from its neighbor's orientation. Thus, a normal piece of iron has no overall magnetic field.

However, if all of the atoms are lined up so that their magnetic fields are in the same direction, the individual magnetic fields will add up to produce an overall magnetic field. Then the piece of iron will act like a magnet.

In the last lesson, you learned that an ordinary nail can be magnetized by placing it next to a permanent magnet. This is because the permanent magnet forces each tiny atomic magnet in the nail to line up. This is just like a compass needle that lines up with a magnetic field. Once many of the atoms have lined up, the nail will be a magnet itself. The more the atoms are lined up, the stronger the nail's magnetic field will be.

Over time, however, the atoms will move out of alignment. This is due to the random motion of the atoms. This motion causes the nail to lose its magnetism. Heating up any magnet will cause it to lose its magnetism. This is because as the magnet's temperature increases, its atoms move around faster. Thus, the atoms move out of alignment more quickly. Similarly, if you drop a magnet, the impact will jostle some of the atoms out of alignment. When this happens, the magnet will lose its magnetism.

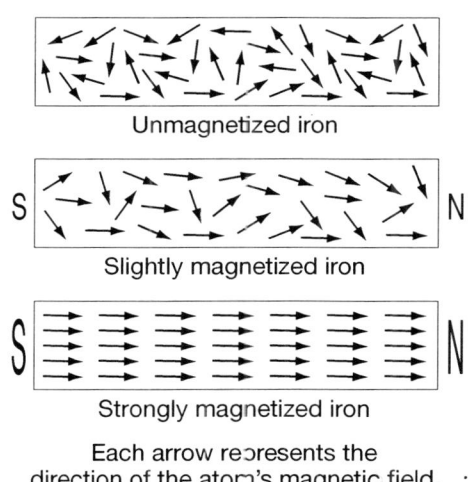

Unmagnetized iron

Slightly magnetized iron

Strongly magnetized iron

Each arrow represents the
direction of the atom's magnetic field.

As we already learned, Earth itself is a giant magnet. However, scientists do not yet completely understand the source of Earth's magnetic field. Earth's center is made up of molten rock, like lava. This molten rock swirls around in the interior of Earth. The molten rock moves charged particles with it. Scientists believe that these moving charged particles may create a magnetic field. This is like charges that move in a wire. Therefore, Earth is a giant electromagnet.

Practice 4—Electric Currents and Magnetism

Directions: Decide whether each statement that follows is true (**T**) or false (**F**). Write the correct letter in each blank.

_____ 1. Electromagnets are the result of moving electric charges, but permanent magnets are not.

_____ 2. The magnetic field that results from a loop of wire is the same as the magnetic field of a bar magnet.

_____ 3. Adding more loops of wire to an electromagnet will increase its strength.

_____ 4. Increasing the amount of current through an electromagnet will increase its strength.

_____ 5. All types of atoms act like a tiny magnet.

_____ 6. You can magnetize a piece of metal by increasing its temperature.

_____ 7. Random vibration of atoms causes some substances to lose their magnetization over a period of time.

_____ 8. Earth's magnetic field most likely results from charged particles moving in the Earth's liquid center.

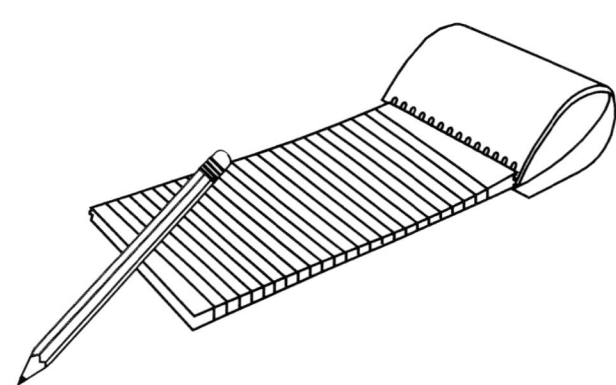

■ LESSON MASTERY TEST

Directions: Match the following electrical quantities with the appropriate unit. Write the correct letter in each blank.

_____ 1. current   (a) volts

_____ 2. electrical energy   (b) amps

_____ 3. resistance   (c) coulombs

_____ 4. charge   (d) ohms

Directions: Circle the answer that correctly completes each of the following statements.

5. The flow of electrons is known as a ____ measures the number of charges flowing and the speed at which they move.

   (a) current   (b) resistance   (c) voltage

6. Increasing the resistance in a circuit ____ the amount of current that flows.

   (a) increases   (b) decreases   (c) does not affect

7. A material that does not let electricity flow through it easily is known as a(n) ____.

   (a) insulator   (b) conductor   (c) superconductor

8. The south poles of two magnets will ____ each other.

   (a) attract   (b) repel   (c) magnetize

9. ____ of electrical energy is needed to cause 2 A of current to flow in a circuit with 3 Ω of resistance. (Remember Ohm's Law: $V = I \times R$)

   (a) 2 V   (b) 6 V   (c) 12 V

10. A circuit that is NOT connected in a complete loop is called a(n) ____.

    (a) open circuit   (b) closed circuit   (c) short circuit

(continued on next page)

Directions: Decide whether each statement that follows is true (**T**) or false (**F**). Write the correct letter in each blank.

_____ 11. Heating a magnet can cause it to lose its magnetization.

_____ 12. All magnets are the result of moving charges.

_____ 13. The motion of protons causes some materials to be magnetic.

_____ 14. Conductors have a large resistance.

_____ 15. If you break a magnet in half, one piece will be a north pole and the other piece will be a south pole.

_____ 16. An electrical outlet supplies electric charges.

_____ 17. Increasing the amount of current through an electromagnet will increase its strength.

_____ 18. In a magnetized bar of iron, most of the atoms are pointing in the same direction.

■ Lesson 2—Applications of Electricity and Magnetism

Goal: To understand several useful applications of electricity and magnetism

Series and Parallel Circuits

If you want to connect two electrical devices in a circuit, such as plugging in two appliances in your home, there are two different ways they can be connected. One way is to put the two devices in the same loop. This means the same electrons flow through both devices. This type of connection is called a **series circuit.** The other way is to make a branch in the circuit. This makes the electrons follow two different paths. This type of connection is known as a **parallel circuit.**

Imagine that you want to connect two lightbulbs to the same battery. You first connect a wire from one end of the battery to the first lightbulb. Then, you connect a wire from the first lightbulb to the second lightbulb. Finally, you connect the second lightbulb to the other end of the battery. When lightbulbs are connected this way, we say that they are connected **in series.** There are no branches in the circuit. There is only one loop. Any electrons that go through the first lightbulb must also pass through the second lightbulb.

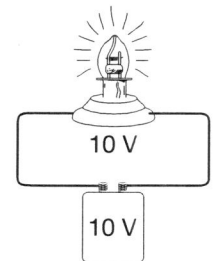

(a) One bulb glows very brightly.

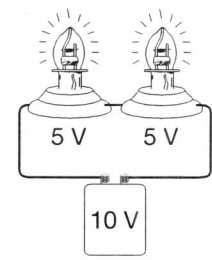

(b) Two bulbs are connected in series. Both glow less brightly.

There is a disadvantage to using the series circuit. The push, or voltage, given to each electron must be split between the two lightbulbs. If 10 V of push, or energy, are given to each electron, then only 5 V can be used in each lightbulb. Thus, less current will flow through each bulb. This means that each lightbulb will not glow as brightly.

On the other hand, you can **branch** the wire. To do this, one loop goes from the battery to the first lightbulb. Another similar loop goes from the battery to the second lightbulb. We then say that the lightbulbs are connected **in parallel.** Now, each electron can go through one lightbulb or the other, but not both. There are two separate paths for the current to follow.

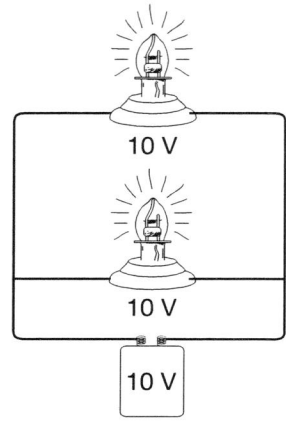

In a parallel circuit, both bulbs glow just as brightly as a single bulb.

Each electron is still given 10 V of energy. However, because each electron only goes through one of the bulbs, all 10 V of energy can be used to push the electrons in that bulb. Thus, the same amount of current flows through each bulb. And, the same amount of current flows when only one bulb is connected. This means each bulb will glow just as brightly as the other. Because each lightbulb has its own path, neither bulb is affected by the other bulb. However, because there are now two bulbs, each with the same amount of current as a single bulb, twice as much current must flow out of the battery to the bulbs.

There are other differences between a series circuit and a parallel circuit. What happens if one lightbulb burns out? The burned out bulb doesn't allow electricity to flow through it anymore. In a series circuit, all the electricity must go through both lightbulbs, so if one isn't working, it will stop the flow to the other one as well. Both lightbulbs will go out. In a parallel circuit, on the other hand, each bulb has its own path to the battery. So, if one bulb burns out, the other one will continue to shine.

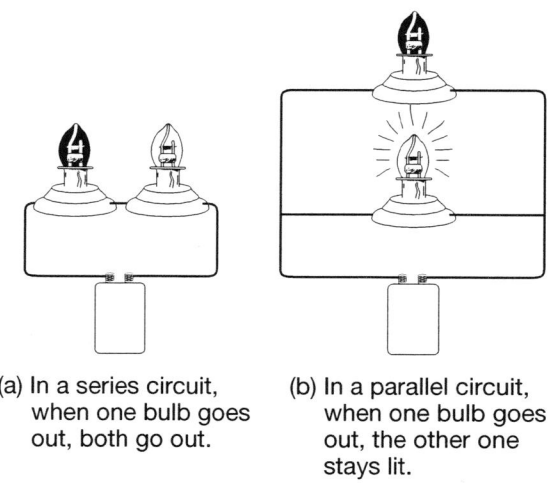

(a) In a series circuit, when one bulb goes out, both go out.

(b) In a parallel circuit, when one bulb goes out, the other one stays lit.

Think About It

Consider strings of holiday lights. Should these lights be connected in series or in parallel?

The outlets in your house allow appliances to be connected in parallel. This is good, because it means that if you turn off a lamp, the television doesn't turn off as well. On the other hand, parallel circuits can be dangerous if they are misused. Each appliance that is added causes more current to flow through the wires in your walls. If too many appliances are used at one time, a very large current can flow in the wires. This can create heat that could lead to a fire. Fortunately, houses have circuit breakers or fuses to protect against this.

Practice 1—Series and Parallel Circuits

Directions: Circle the answer that correctly completes each of the following statements.

1. The lightbulbs at right are connected ____.

   (a) in series   (b) in parallel

2. If two appliances are connected in series, ____ current will flow in each one than if each one was connected separately.

   (a) more   (b) less

3. If two appliances are connected in parallel, and one stops working, the other one ____.

   (a) will stop also   (b) will continue to work

4. A battery must supply ____ current when two lightbulbs are connected in parallel than when they are connected in series.

   (a) more   (b) less

5. The outlets in your house ____ appliances to be connected in parallel.

   (a) allow   (b) do not allow

The Motor Effect

In addition to attracting other magnets, a magnet can also exert a force on individual charged particles, such as electrons. For example, a single electron moving near a magnet will be deflected by the magnet. The direction the electron is pushed by the magnet is always perpendicular to the direction the electron is moving.

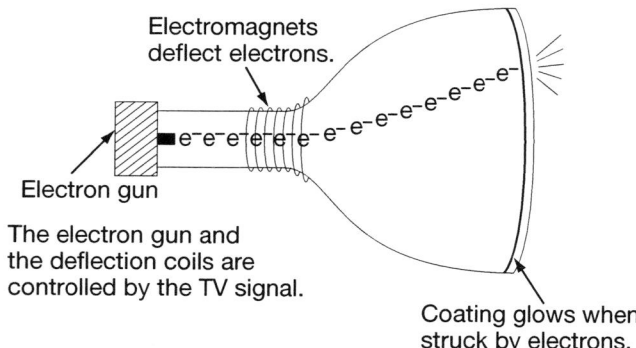

Television Picture Tube

This principle is used in televisions. Inside the television is a device that fires electrons at the back of the glass television screen. The back of the screen is covered with a substance that glows when electrons hit it. The more electrons that are fired, the greater the glow. The television image is formed by aiming the electrons in a zig-zag motion from the top of the screen to the bottom.

The electrons are aimed by using a pair of electromagnets. The amount of current through the electromagnets is controlled by the signal received from the television station. The amount of current through the magnets determines how strong their magnetic field is. This determines how much the electrons will bend. The signal from the television station steers the electrons to the proper place, making an image on the screen. The entire picture is scanned out by electrons thirty times every second.

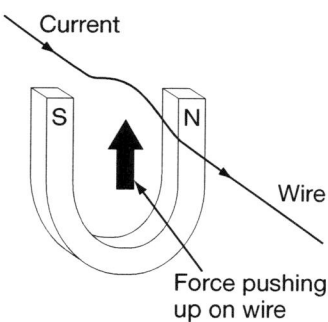

A current (moving charges) in a wire feels a force.

Because a current in a wire is made up of many moving electrons, a wire carrying current will also feel a force near a magnet. The more current there is flowing in the wire, the greater the force it will feel. This relationship is known as the **motor effect.** The motor effect is responsible for almost all devices which use electricity to create motion.

One of the earliest applications of the motor effect was in the ammeter. An **ammeter** is a device for measuring the amount of current in a wire. An ammeter consists of a coil of wire between two magnets. When current flows in the wire, the magnets exert a force on the wire, which causes the wire to twist. The more current there is flowing through the coil, the more force it will feel, and the further it will turn. A needle is attached to the coil. A calibrated scale behind the needle allows the user to determine how much current a certain deflection indicates.

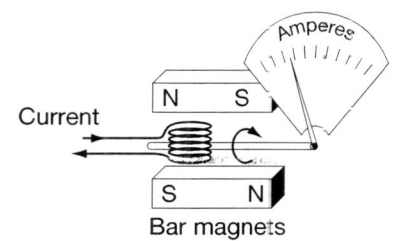

Ammeter

An electric motor functions similarly to an ammeter. However, this time, the coil is allowed to rotate continuously. The magnet creates a force on the coil, pushing the coil to line up with the magnet. Once it is lined up, however, the current flowing through the coil is reversed. Then, the coil is pushed to line up in the opposite way. Thus, the coil continues rotating until it is lined up in the opposite direction. Then, the current is reversed again. This cycle repeats, creating a continuous rotation of the coil. This rotation can be used to drive the wheels of an electric car or to rotate the blades of an electric mixer.

Adding extra loops to the coil creates a stronger force, making a stronger motor. Similarly, stronger magnets on the outside will make a more powerful motor. Also, because the motor described above only has two magnets, it will feel a strong push only twice during each rotation.

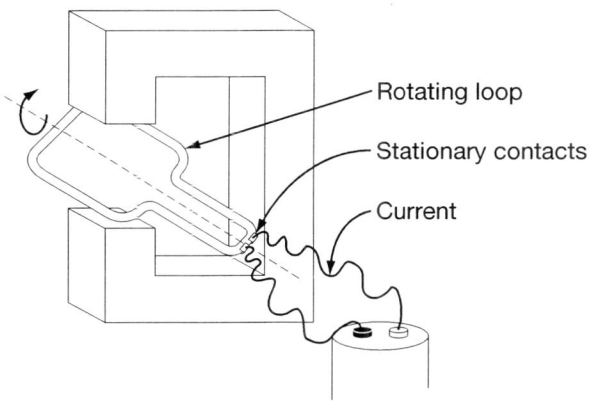

A Simple Motor

Practice 2—The Motor Effect

Directions: Decide whether each statement that follows is true (**T**) or false (**F**). Write the correct letter in each blank.

_____ 1. A magnet can only exert a force on other magnets.

_____ 2. A television uses gravitational force to aim electric charges.

_____ 3. The more current there is in a wire, the more force it will feel from a magnet.

_____ 4. An ammeter is used to measure magnetic fields.

_____ 5. In a motor, the coil turns due to the electric force.

_____ 6. In a motor, the current must switch directions in order for the coil to turn continuously.

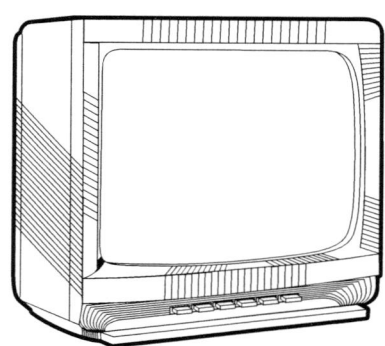

The Generator Effect

In the last section, you learned that a magnet can cause a wire carrying current to move. It turns out the opposite can happen as well. A magnet can create a current in a moving wire. This is known as the **generator effect.**

As you learned in Lesson 1, when you move a wire near a magnet, the charges in the wire feel a force. This means that both the protons in the nucleus and the loose electrons feel a force. However, the protons are held tightly in the nucleus, so they cannot move. Only the electrons are free to move. Thus, the electrons will be pushed along the wire. In other words, a current flows.

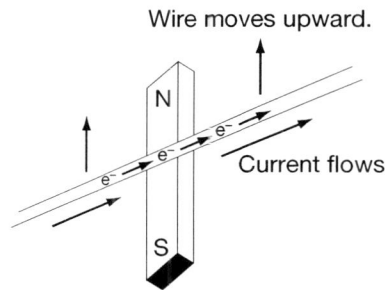

Wire Moving Near a Magnet

The push created by the magnetic field is exactly the same as the push caused by a battery. In fact, the push *is* a voltage. The voltage created will be stronger if the wire is moved more quickly, or if the magnet is stronger.

The most important application of this is in the electric generator. In a generator, a large coil of wire is rotated between several large magnets. As the wire turns, the magnetic field pushes on the electrons. This push creates a voltage in the wire, which can be used to run any electric appliance. In fact, power stations use this same procedure to create the electric power that is used in our homes.

In the generator, the coil of wire is generally attached to a large shaft, known as a **turbine.** Some source of mechanical energy is necessary to cause the turbine to rotate. Most power stations use the pressure of steam to push blades on the end of the turbine. The steam can be created by burning coal or oil to heat water, or by using the heat from a nuclear reaction.

It is important to remember that the generator does not *create* energy. It simply changes the mechanical energy of the rotating coil into the electric energy of the voltage pushing the electrons. The energy must initially come from some other source, such as the chemical energy of fossil fuels.

When the coil in a generator rotates, it produces a current that is constantly changing directions. As the coil goes through one rotation, it is first turning toward the north pole of the magnets, and then turning towards the south pole of the magnets. As it moves toward the north pole, the electrons are pushed one direction, and as it moves towards the south pole, the electrons are pushed in the opposite direction. So, the current changes direction twice per revolution. This is known as an **alternating current.** A battery, on the other hand, creates a **direct current.** This is because a battery always pushes the electrons in the same direction. The voltage supplied in an electrical outlet is an alternating current. The direction of current in an electrical outlet changes 60 times per second.

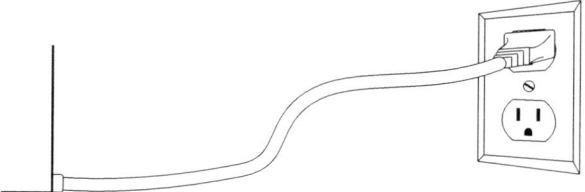

In addition to a wire moving near a magnet, a current can be created by moving a magnet near a wire. There is no difference between the magnet moving near the wire and the wire moving near the magnet. The same voltage results. Generators could be created by spinning magnets around a wire, but this would be much more difficult than spinning the wire.

Practice 3—The Generator Effect

Directions: Decide whether each statement that follows is true (**T**) or false (**F**). Write the correct letter in each blank.

_____ 1. When a wire moves near a magnet, a resistance can be created in the wire.

_____ 2. The faster a wire moves near a magnet, the more voltage is created in the wire.

_____ 3. An electric generator creates energy that did not exist before.

_____ 4. Using stronger magnets in a generator will allow more voltage to be produced.

_____ 5. Alternating current always flows in the same direction.

_____ 6. A battery supplies direct current.

_____ 7. It is possible to create a current in a wire by moving the wire near a magnet, but not by moving the magnet.

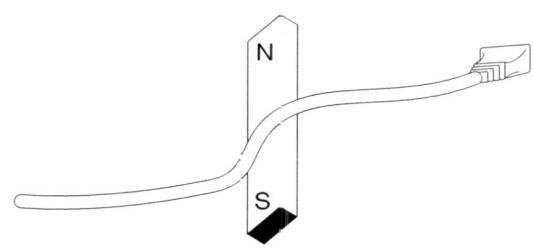

Recording and Creating Sound

Sound is a wave traveling in air molecules. Sound can be recorded electrically by using a voltage to represent the sound wave. A microphone is used to convert sound waves to a voltage. For example, when the air molecules are pushed close together, the voltage is made to be high. When the air molecules are spread apart, the voltage is made to be low. The sound can be recreated by applying this changing voltage to a loudspeaker to reproduce the motion of the air molecules.

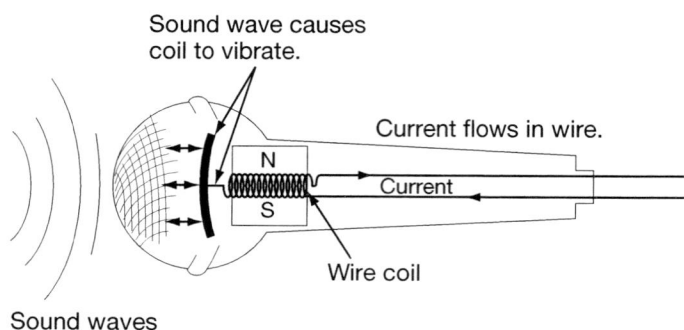

Microphone

A microphone is used to convert sound to a voltage signal. And, a loudspeaker is used to convert the voltage signal back to sound waves. There are several ways microphones can do this. One type of microphone converts sound into an electric signal by using the generator effect. This type of microphone has a small coil of wire and a magnet. When a sound wave hits the microphone, it causes the coil to vibrate back and forth. This vibration follows the pushing and spreading of the air molecules in the sound wave. The motion of the coil of wire near the magnet creates a voltage in the wire. This voltage corresponds to the vibration of the air in the sound wave.

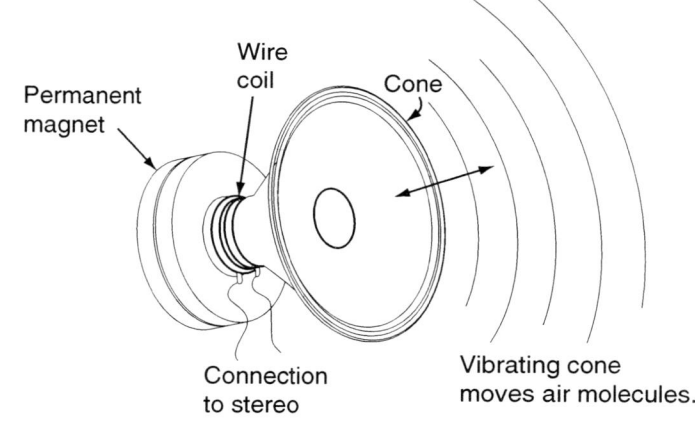

Loudspeaker

A loudspeaker generates sound using the reverse process. Like the microphone described above, the speaker has a coil of wire and a magnet. The changing voltage that represents a sound wave is connected to the wire. The coil acts like an electromagnet. The electromagnet then attracts or repels the permanent magnet. As the voltage varies, the coil moves back and forth, vibrating in the same way as the original sound wave. The coil is attached to a large paper or cloth cone. The cone pushes or pulls on the air, creating a sound wave identical to the one that was recorded.

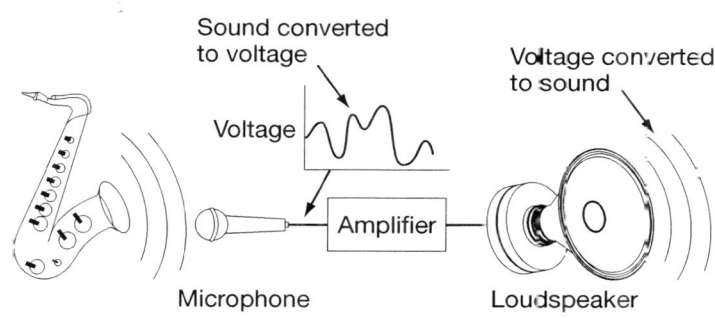

If you look inside a broken speaker, you can see these parts. On the back of the paper cone of the speaker is a coil of wire. This coil is connected to the wires which carry the electric signal from the stereo. Around this coil is a circular magnet, which will attract paper clips and any other metal object. Even if you can't open up a speaker, you can see the effect of this magnet by holding a compass near the speaker. The needle will swing around to point toward the magnet. This is also why "home theater" speakers that are placed close to the TV must shield the speaker's magnet and coil. Otherwise, the speaker can interfere with the electrons in the TV that are aimed to make the picture.

After a sound is converted into a voltage, there must be a way to preserve this signal until it is played back by a speaker. There are three common techniques to store this signal. They are vinyl records, cassette tapes, and compact discs (CDs).

On a record, a circular path is etched into the vinyl. Although the circular lines look straight, if you look very closely you will see they are actually wavy. The shape of the wave corresponds to the pattern of the sound wave. When the needle of the record player moves along this path, the needle vibrates along with the wave. This vibration is converted into an electric signal, which is used to drive a speaker to create the sound.

With a tape recorder, the electric signal is converted into a magnetic pattern on the moving tape. When recording, the electric signal is sent to a tiny electromagnet in the recording head that touches the tape. This creates a magnetic field that changes according to the electric signal. The tape inside the cassette is actually covered with a metallic coating, which can be magnetized by the electromagnet in the recording head. This leaves a pattern of magnetization on the tape as it moves inside the cassette. This magnetic pattern corresponds to the sound wave.

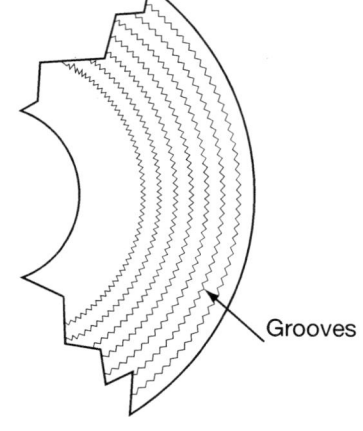

Grooves

Wavy lines vibrate a needle riding in the grooves.

Vinyl Record

When playing a cassette, the generator effect is used. The play head consists of a small coil of wire. As the magnetized tape moves by, the generator effect causes a small current to be created in the wire. This current corresponds to the pattern of magnetization on the tape, which represents the original sound wave.

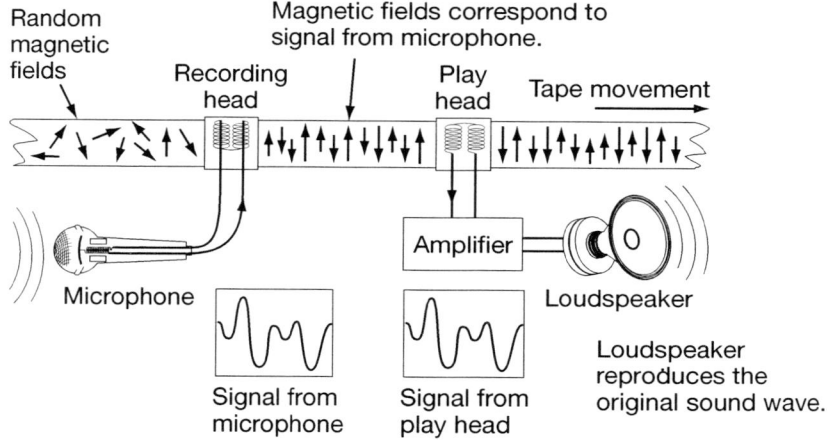

Tape Recorder

Finally, in a CD, the electric signal representing the sound wave is converted mathematically into a series of 1's and 0's. The compact disc has a reflective surface, but in the recording process, tiny pits are made into the surface. At these points, light does not reflect. The series of reflective and nonreflective points represents the 1's and 0's of the electric signal.

The CD is played by shining a small laser onto the CD as it spins. As the laser passes each point, a sensor detects whether the light is reflected at that point or not. If the light is reflected, a voltage is generated. The pattern of 1's and 0's are then mathematically converted back to the original sound waves.

Records can wear out quickly, because any physical scratch on the surface will disrupt the sound wave that is produced. Furthermore, as the needle wears out, it can wear out the record. A cassette tape will eventually wear out, too. This is because the tape slowly becomes demagnetized. This can be a result of other magnets near the cassette (holding a magnet next to the cassette can ruin it). Also, the gradual random vibrations of the magnetized atoms can demagnetize the tape. A CD can last a long time without wearing out. This is because light is the only thing that touches the CD surface when it is played. Therefore, there is no damage to the CD.

Practice 4—Recording and Creating Sound

Directions: Decide whether each statement that follows is true (**T**) or false (**F**). Write the correct letter in each blank.

_____ 1. It is possible to represent a sound wave using an electrical voltage.

_____ 2. A microphone converts sound into electricity using the generator effect.

_____ 3. A speaker uses an electromagnet to create electromagnetic waves.

_____ 4. A speaker converts electrical energy into sound energy.

Directions: Decide whether each process below describes a record (**R**), cassette (**C**), or compact disc (**D**). Write the correct letter in each blank.

_____ 5. The reflection of laser light is used to determine a pattern of 1's and 0's, which corresponds to the sound wave.

_____ 6. A thin metallic layer is magnetized to correspond with the voltage signal.

_____ 7. The vibrations of a needle are converted into electrical signals.

_____ 8. The generator effect of a moving magnetized strip causes current to flow in a small coil.

■ LESSON MASTERY TEST

Directions: Look at the list of terms below. Write the letter of the correct term in the blank before each description. (*Hint:* You will not use all of the choices.)

(a) electric motor   (d) microphone   (g) compact disc

(b) electric generator   (e) cassette tape   (h) ammeter

(c) speaker   (f) record   (i) television

_____ 1. A coil of wire is placed between two magnets. The more current that flows in the wire, the more the coil will twist.

_____ 2. As a magnetized tape moves past a coil of wire, it creates a voltage in the coil.

_____ 3. A coil of wire is forced to spin in between two magnets, creating a voltage in the coil.

_____ 4. Laser light hits either a reflective or nonreflective point.

_____ 5. A current runs through a coil of wire, but the current changes direction so that the coil spins continuously.

_____ 6. Sound waves hit a coil of wire, which moves back and forth in front of a magnet, creating a voltage in the coil.

_____ 7. The path of electrons is controlled using electromagnets.

Directions: Circle the answer that correctly completes each of the following statements.

8. The lightbulbs at right are connected _____.

   (a) in series   (b) in parallel   (c) alternately

9. A generator creates _____ current.

   (a) alternating   (b) direct   (c) alternative

10. To create a greater voltage in a wire near a magnet, you must move the wire _____.

   (a) faster   (b) slower   (c) for a longer time

(continued on next page)

Directions: Decide whether each statement that follows is true (**T**) or false (**F**). Write the correct letter in each blank.

_____ 11. It is possible to represent a sound wave using an electrical voltage.

_____ 12. Using stronger magnets in a generator will allow more voltage to be produced.

_____ 13. An electric generator creates electric charges.

_____ 14. A magnet can exert a force on moving electric charges.

_____ 15. The outlets in your house allow appliances to be connected in parallel.

_____ 16. If two lights are connected in series, and one stops working, the other one will stop also.

■ Lesson 3—Modern Physics

Goals: To understand the developments in physics over the last century; to learn about some of the mysteries of the universe which remain unsolved

Relativity

In 1905, German scientist Albert Einstein published the **Special Theory of Relativity.** This described strange phenomena that occur when an object moves close to the speed of light. He based his theory on experiments concerning the speed of light.

To understand Einstein's theory, you first must understand the idea of relative motion. **Relative motion** is how the motion of one moving object appears to another moving object. Suppose you are driving in a car at 30 mph. You approach a bicyclist moving at 20 mph. The time it takes for you catch up to him is the same as if he were standing still and you were going 10 mph. The difference between your speed and that of the bicyclist is 10 mph. The difference between the speed of the car and the speed of the bicycle is known as the **relative speed.** On the other hand, suppose you are going in the opposite direction as the bicyclist. Then your relative speed would be 50 mph.

Imagine that you are in a train that is moving at 60 mph. If you throw a baseball up in the air, it will go up and come back down, just as if the train were standing still. In fact, if there were no windows, and the track was very smooth, you would have no way of knowing that the train was moving.

50 mph relative speed

20 mph

30 mph

30 mph

10 mph relative speed

Relative Speed

Suppose you throw the ball down the aisle to your friend at a speed of 40 mph. Because the ball is moving *and* the train is moving, to a person on the ground looking in the windows, the ball will appear to be moving at 100 mph. The relative speed, or relative velocity, of the ball is the speed the train is moving plus the speed the ball is moving (60 mph + 40 mph = 100 mph).

Now, suppose, instead of throwing a ball, you turn on a flashlight. Of course, the speed of light is too fast to see. But, if you could measure it, the beam would move away from you at the speed of light (186,000 miles per second). You might expect that the person on the ground would see the light beam moving away at that speed *plus* the 60 mph speed of the train.

However, careful experiments were made at the end of the 19th century. These experiments showed that the person on the ground sees the light beam moving at the *same* speed as the person in the train sees it! **The speed of light is always the same,** no matter how fast the person who sees it is moving. This is as if the bicyclist appeared to be moving at 20 mph no matter how fast you passed him. Or, it is as if the baseball didn't move any faster when thrown on a moving train.

Think About It

Imagine you are riding in a spaceship at ½ the speed of light. If you turn on the headlights, how fast will the beam move away from you? How fast will the light seem to be traveling to someone in a stationary spaceship that you pass?

This amazing observation is one of the two rules at the base of Einstein's theory of relativity. The other rule states that all motion is relative. To understand this, think back to the example of the ball in the moving train. If there are no windows, everything appears the same inside the train as it would if the train were standing still. In fact, the scenery could be moving while the train is standing still. According to Einstein, we cannot say that one object is stationary while another object is moving. We can only say that two objects are moving compared to, or **relative to,** each other.

This is because there is no stationary point in the universe against which to compare all other movement. You may say that a person on the ground is standing still. But, in fact, the Earth is moving around the sun. And, the sun is moving around the galaxy, and so on. There is nothing in the laws of physics that will allow you to say that something is absolutely at rest.

Thus, the two rules of Special Relativity are:

1. The speed of light is always the same, regardless of the speed of the observer.

2. All motion is relative. According to the laws of physics, we can only say that two objects are moving compared to each other. We cannot say that an object is definitely at rest.

With some difficult mathematical manipulation, Einstein derived some amazing consequences of these two rules. One consequence is that as objects move faster, time actually slows down for them.

If the observer could look inside a spaceship that was moving past him at close to the speed of light, everything inside would be happening slowly. For example, the clocks would move more slowly, a ball would fall more slowly, and the people would even age more slowly. This is because their time passes at a different rate. Furthermore, the same spaceship would be compressed in the direction it is moving. Just as the length of time is decreased, physical length is also decreased. A ruler that is 1 foot on Earth would appear to the observer to be only 6 inches long!

However, each of these changes are relative. The person standing in the ship sees nothing unusual. Because the spaceship isn't moving relative to him, nothing changes. The clocks move normally, and his ruler doesn't change length.

Furthermore, suppose you were to get into a spaceship and travel close to the speed of light to another star, and come back. You may think the journey took 5 years, but when you come back, 10 years will have passed on Earth. Your time slowed down, but you didn't notice it.

This may seem strange, but it has actually been tested with modern clocks that are extremely accurate. Scientists used two clocks which run exactly the same. One clock traveled around the Earth in a plane, while the other remained stationary on the ground. When the clock was brought down from the plane, it was actually a fraction of a second behind. This indicates that time had slowed down slightly while it was in motion.

These effects become greater the faster an object is moving. At the speed of everyday objects, the effects of relativity are far too small to be observed. Only when an object approaches the speed of light would the effects of relativity become noticeable.

One consequence of the theory of relativity is that it is impossible for an object to move faster than the speed of light. As a result of relativity, once an object moves faster than light, its time would stop. This means it would shrink out of existence. Thus, no object, not even some future spaceship, can move faster than light.

Practice 1—Relativity

Directions: Decide whether each statement that follows is true (**T**) or false (**F**). Write the correct letter in each blank.

_____ 1. The rules of Special Relativity are easily observed in everyday objects.

_____ 2. If you are riding a bicycle at 20 mph, and you pass a person running at 7 mph in the same direction, your relative speed is 13 mph to the person running.

_____ 3. If you are riding a bicycle to deliver newspapers at 20 mph, and you can throw the newspapers straight ahead at 15 mph, the total speed of the newspapers relative to the ground is 35 mph.

_____ 4. The speed of light is the same, no matter how fast you are moving relative to the light.

_____ 5. It is impossible to say that an object is absolutely stationary, because there is nothing stationary in the universe to compare it to.

_____ 6. If you could observe a clock in a spaceship traveling close to the speed of light, it would be running much faster than an ordinary clock.

_____ 7. Objects become stretched out when traveling close to the speed of light.

_____ 8. If you were traveling close to the speed of light, you would feel as if everything around you was happening very slowly.

_____ 9. If an object has enough energy, it can go faster than the speed of light.

Quantum Mechanics

The theory of relativity explains some strange consequences of moving very fast. The theory of **quantum mechanics** predicts some strange consequences about objects that are extremely small.

In *Sound and Light,* you learned that light comes in small bundles called photons. Each photon is the smallest amount of light that can exist. The smallest amount of any substance is known as a **quantum.** For example, 1 cent is the quantum for United States money. Any price is some multiple of 1 cent. For a bar of iron, the quantum is one iron atom.

Quantum mechanics predicts that many other things in nature always come in multiples, or quanta, like the examples above. For example, the energy of an object must always be a multiple of some small amount of energy. Even the speed of an object is a multiple of some tiny fundamental speed.

We never notice this effect, known as **quantization.** This is because the fundamental quanta are so small. However, imagine what would happen if the quantum for length, for example, was 1 inch. As you grew, your height would not change continuously. For one year, you might be 5 feet, and then suddenly you would jump to 5 feet, 1 inch. Or, if the quantum of speed was 5 mph, when driving a car you would start from rest, then suddenly jump to 5 mph, then jump to 10 mph. It would be impossible to move at any speed in between.

At very small sizes, however, such as the size of the atom, quantum effects are very obvious. One effect you learned about earlier was the orbits of electrons. Electrons can only move in certain paths, determined by the fundamental quantum of energy. It is physically impossible for an electron to orbit in between these paths. This is similar to running around a track, and finding it impossible to run in between the lanes.

There is another strange effect of quantum mechanics you learned. This is wave-particle duality. Very small objects tend to act like both waves *and* particles. For example, if a particle like an electron is sent through a narrow opening, its path tends to spread out on the other side, just like a wave would. Imagine if a baseball going through a window began to spread out as it passed through!

One of the most surprising and controversial discoveries of quantum mechanics is about extremely small particles. The motion of an extremely small object can only be predicted as a probability. For example, when an electron is pushed into a higher orbit, it can spontaneously drop back down to a lower orbit, putting out a photon of light. However, if you were watching the particle, it would be impossible to predict when the photon would be emitted. You could only say, for instance, "There is a 50% probability that it will fall in the next minute." This is not due to lack of human knowledge. It simply is due to the fact that particles obey probabilities rather than definite rules.

Imagine if everyday objects behaved this same way. If you set up a line of dominoes, they could spontaneously fall over. This would happen even without a push of any kind. However, it would be impossible to predict when the dominoes would fall. Somehow, the dominoes would simply fall. Therefore, if quantum mechanics were applied to everyday objects, the motion of all objects would be as random as the rolling of dice.

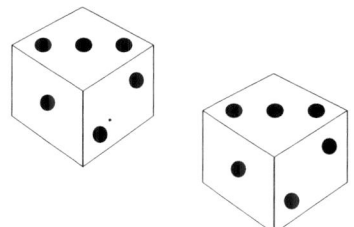

If you find it difficult to believe these discoveries, don't feel bad. Even Albert Einstein could not accept the ideas of quantum mechanics. He once stated, "I cannot believe that God plays dice with the universe." This statement showed his reluctance to believe that the future of the universe is determined by chance.

Practice 2—Quantum Mechanics

Directions: Decide whether each statement that follows is true (**T**) or false (**F**). Write the correct letter in each blank.

_____ 1. The effects of quantum mechanics are most visible in objects smaller than atoms.

_____ 2. A photon is the quantum of light.

_____ 3. We can observe the effect of quantization of speed when driving a car.

_____ 4. An electron can choose any path to orbit around the nucleus.

_____ 5. All very small particles sometimes behave like waves.

_____ 6. The outcome of interactions between sub-atomic particles is based on probability, and cannot be predicted.

Radioactivity

When two atoms come together in a chemical reaction, only the electrons on the outside of the atom are involved in the reaction. The nucleus of each atom does not change. However, radioactivity and nuclear reactions involve changes in the nuclei of atoms.

Remember that the number of protons in the nucleus (known as the **atomic number**) determines the type of element. The number of neutrons in a certain element can vary somewhat. Atoms with the same number of electrons but different numbers of neutrons are called **isotopes.**

There are two forces responsible for holding the nucleus together. The force that holds the protons and neutrons to each other is known as the **strong force.** Furthermore, protons and neutrons are each made of smaller particles known as **quarks,** which are held together by the **weak force.**

In some types of atoms, these forces are not strong enough to hold the nucleus together permanently. Over time, the nucleus can decay, breaking into smaller parts. These are known as **radioactive** nuclei.

When a nucleus decays radioactively, there are three different types of particles that are emitted. These particles are alpha rays, beta rays, and gamma rays. An **alpha (α) ray** consists of particles made of two protons and two neutrons bonded together. Alpha rays are emitted when the strong force is overcome, and two protons and two neutrons break off of the nucleus. A **beta (β) ray** consists of electrons, which are emitted from the nucleus when the weak force allows a neutron to break apart. **Gamma (γ) rays** are a form of electromagnetic waves, just like light or X-rays, but with much higher energy.

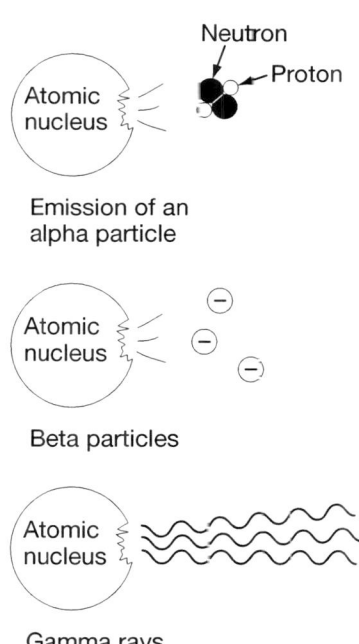

Emission of an alpha particle

Beta particles

Gamma rays

These three types of radiation have different abilities to move through material. This ability is known as their **penetrating power.** Alpha rays consist of big particles, so they move relatively slowly. Thus, they can be stopped the easiest. A sheet of cardboard would be enough to protect you from a radioactive substance emitting alpha rays only. Beta particles are lighter, and move faster. Thus, they can penetrate paper, but are easily stopped by a thin sheet of metal, such as aluminum foil. Gamma rays have the most penetrating power. It requires a very thick layer of dense metal, such as lead, to stop gamma rays.

Radioactive source

Alpha (α) particles can be stopped by paper.

Paper

Beta (β) particles can be stopped by aluminum foil, but penetrate paper.

Aluminum foil

Gamma (γ) rays can be stopped by a thick layer of lead, but penetrate paper and aluminum foil.

Lead

Penetrating Power

Radioactive decay involves changes in the nucleus. Therefore, the atom changes to a different type after a radioactive decay. For example, in alpha decay, the nucleus loses two protons, so its atomic number decreases by two. Thus, a radioactive element that undergoes alpha decay will change into an element with an atomic number of two less than it was before the decay.

Some types of atoms, such as those of uranium, undergo many consecutive radioactive decays before they are stable. And, some isotopes of uranium are unstable. One isotope undergoes 9 alpha decays and 9 beta decays. When the decay process is finished, the isotope ends up as a completely different element. This element is lead. There are many other examples of radioactive decay. They all involve a radioactive substance that changes into another type of element.

Scientists can describe how quickly a radioactive substance decays. They do this by defining the half-life of a substance. The **half-life** of a radioactive substance is the length of time it takes for half of the radioactive substance to decay. For example, one isotope of carbon, known as carbon-14, can decay into nitrogen. Its half-life is about 6,000 years. This means if you have one pound of carbon-14, after 6,000 years it will be ½ pound of carbon-14 and ½ pound of nitrogen. After 12,000 years, another half of the remaining carbon-14 decays. This leaves ¼ pound of carbon-14.

A substance with a very short half-life decays very quickly. This means it will put out a lot of radiation at first. This is because many atoms are decaying. However, the material will also quickly turn into a nonradioactive substance. A long half-life means it is not putting out much radiation, but will remain radioactive for a much longer time.

Radiation is very dangerous to living things. This is because of the damage caused when the rays hit cells in the body. Some particles directly destroy the cells when they collide. Other particles damage the cell's DNA, molecules which control all of the processes in the cell. This disrupts the normal function of the cells, leading to diseases such as cancer. Furthermore, the damage to the DNA can be passed on to future generations.

However, radioactivity has many useful applications as well. One is **radioactive dating.** This allows scientists to determine the age of very old objects. For example, all living things contain a small amount of carbon-14. When they die, the carbon-14 decays into nitrogen. By measuring the amount of carbon-14 left in a fossil, for example, scientists

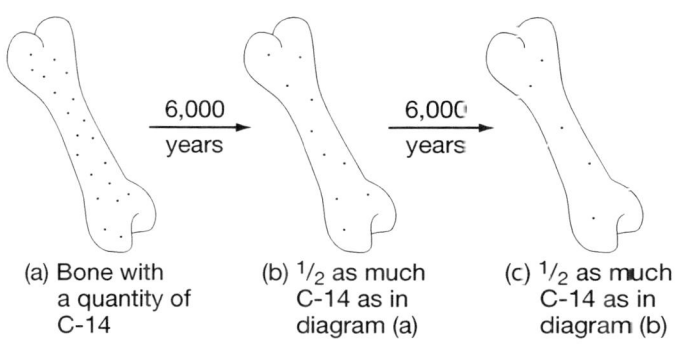

(a) Bone with a quantity of C-14

(b) ½ as much C-14 as in diagram (a)

(c) ½ as much C-14 as in diagram (b)

can determine old it is. If only half of the carbon-14 is left, then the fossil is 6,000 years old (the half-life of carbon-14.) If only one quarter of the carbon-14 is left, then the fossil is 12,000 years old.

Radiation is also very useful in medicine. For example, cancer cells in a cancerous tumor begin to grow very rapidly. It turns out that tumor cells are more easily damaged by radiation than normal cells are. Thus, radiation can be used to kill tumor cells, helping to cure a cancer patient.

Radioactive tracing is another useful application of radioactivity. It takes advantage of the fact that radioactive isotopes act just like their nonradioactive isotopes. However, the radioactive isotopes can be detected by measuring the radiation they emit. For example, a doctor can study a patient's digestion by having the patient eat food with a tiny (and safe) amount of a radioactive isotope. This small amount of isotope is known as a **tracer.** Later, the doctor can measure the amount of radiation emitted at different parts of the body. This allows the doctor to observe how the food was processed by the body.

Practice 3—Radioactivity

Directions: Circle the answer that correctly completes each of the following statements.

1. Chemical reactions only involve interaction between the _____ of atoms.

   (a) electrons   (b) protons   (c) nuclei

2. The _____ force binds protons and neutrons together in the nucleus.

   (a) electric   (b) weak   (c) strong

3. _____ rays consist of 2 protons and 2 neutrons.

   (a) Alpha   (b) Beta   (c) Gamma

4. _____ rays are a form of electromagnetic radiation.

   (a) Alpha   (b) Beta   (c) Gamma

5. _____ rays have the greatest penetrating power.

   (a) Alpha   (b) Beta   (c) Gamma

6. If a radioactive substance has a half-life of 100 years, _____ of it will be left after 200 years.

   (a) all   (b) 1/2   (c) 1/4

Cosmology and the Big Bang

Quantum mechanics and radioactivity answer questions about very small objects. **Cosmology** is the study of the largest objects, such as stars, galaxies, and the entire universe.

Astronomers have many ways to study objects in outer space. In addition to looking at the light they produce, they can also study other forms of electromagnetic radiation. These can be radio waves, infrared waves, X-rays, and even gamma rays. All of these types of radiation are emitted by stars. Each type of radiation can tell scientists different information.

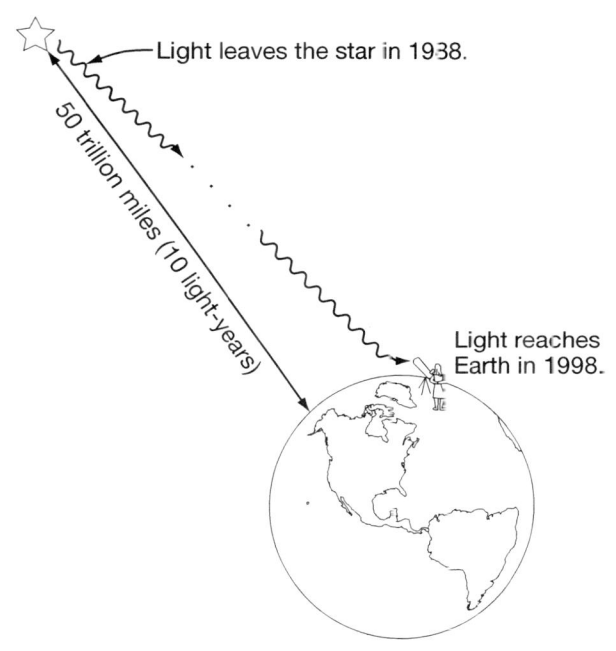

Light leaves the star in 1938.

50 trillion miles (10 light-years)

Light reaches Earth in 1998.

Although light seems to travel instantaneously on Earth, the speed of light is actually slow compared to the distances between stars and galaxies. For example, the nearest star to our solar system, Alpha Centauri, is actually 24 trillion miles away. Because these distances are so large, scientists often measure them in **light-years.** One light-year is the distance light can travel in one year, about 5 trillion miles. So, Alpha Centauri is 4.5 light-years away. This means it takes light 4.5 years to travel from Alpha Centauri to Earth. And that's the nearest star! The light from the most distant objects we can see takes billions of years to reach Earth.

Because light takes so long to travel these distances, you are actually looking into the past when you see the light from stars. The light from a star 10 light-years away left the star 10 years ago. So, you are seeing the star the way it appeared 10 years ago. The light from a star a billion light-years away left the star a billion years ago. So, you are actually looking a billion years into the past. Thus, by looking at very distant objects, scientists can discover how the universe has changed over time.

One of the first discoveries scientists made about the history of the universe is that the entire universe is actually expanding. When scientists look at distant stars, the spectrum of the light the stars emit is lower in frequency than expected. They realized that this is due to the Doppler effect. Remember, the Doppler effect causes waves from an object that is moving away from you to be shifted to a lower frequency. This is like the siren of an ambulance that is moving away from you. Scientists realized that, over time, other stars and galaxies are moving away from us.

Scientists concluded that, in fact, the whole universe is expanding. By tracing this expansion backward, they have calculated that the universe began with all of its matter concentrated at one point. The universe began about 15 billion years ago. This tiny point exploded, pushing out all of the matter. This explosion is known as the **big bang.** Ever since the big bang, the universe has continued to expand outwards as a result of this explosion.

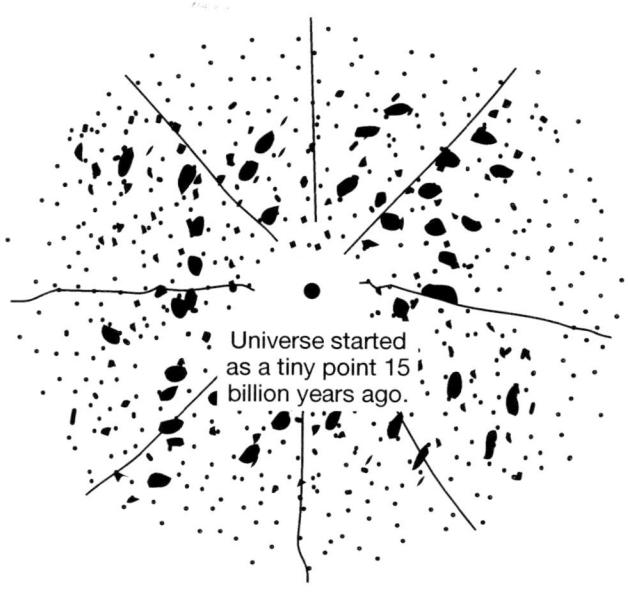

Universe started as a tiny point 15 billion years ago.

The Big Bang

So, will the universe continue to expand outward forever? Scientists are still not sure. The only force that could stop the universe from expanding is gravity. If there is enough mass in the universe, then the force of gravity between all of the objects will be enough to stop it and pull it back in. In fact, the universe could actually collapse back in on itself, just like a baseball thrown upward turns around at the top of its path and comes back down.

If this happens, the universe could crunch back down into a tiny point again. Scientists speculate that this could cause another big bang. Thus, the expansion and collapse of the universe could be a cycle which could repeat over and over again. In fact, it may be that our universe began when a previous universe collapsed!

However, whether the universe will expand forever, or turn around and collapse, depends on how much mass there is in the universe. Therefore, scientists have tried to estimate the total mass of all the objects in the universe. There are two ways to do this. One way is to add the mass of all the objects we can see or detect using special instruments. This is known as the **visible matter.** The other way is to look at the force of gravity on other galaxies and stars. This tells us how much matter there must be pulling on them. However, the visible matter is only 10% as much as the matter predicted by gravitation.

This is a big mystery for scientists. It means that either there is a lot of matter that we cannot yet detect, or there is some unknown force affecting the motion of stars and galaxies. Most scientists believe the first hypothesis. The search for this invisible matter, or **dark matter,** as it is known, is one of the leading areas of research in astronomy today.

Practice 4—Cosmology and the Big Bang

Directions: Decide whether each statement that follows is true (**T**) or false (**F**). Write the correct letter in each blank.

_____ 1. It takes less than a second for light from other stars to reach Earth.

_____ 2. The distance light travels in one year, 5 trillion miles, is known as one light-year.

_____ 3. If a star is 1 million light-years away, we are looking back in time 1 million years when we see it.

_____ 4. Astronomers can observe many different types of electromagnetic radiation from stars.

_____ 5. Scientists have discovered that other stars and galaxies are all moving toward Earth.

_____ 6. The big bang theory states that the universe started at one tiny point, which exploded outward and is continuing to expand.

_____ 7. If the universe contains too much matter, it will continue to expand outward forever.

_____ 8. Scientists believe that they can see all of the types of matter that are in the universe.

■ LESSON MASTERY TEST

Directions: Decide whether each phrase below describes relativity (**R**), quantum mechanics (**Q**), radioactivity (**RA**), or cosmology (**C**). Write the correct letter(s) in the blanks.

_____ 1. The universe was created in the big bang.

_____ 2. Objects that move close to the speed of light become shorter in one direction.

_____ 3. Small objects act as both particles and waves.

_____ 4. These are laws of physics that follow probability only.

_____ 5. Either alpha, beta, or gamma rays can be emitted from the nucleus.

_____ 6. It is impossible to say that one object is absolutely at rest. We can only compare motion between two objects.

_____ 7. Scientists believe that there is some form of matter that they cannot yet detect.

_____ 8. One element can change into another type of element, such as carbon changing into nitrogen.

Directions: Circle the answer that correctly completes each of the following statements.

9. Radioactive decay involves the _____ of an atom.

   (a) electrons   (b) nucleus   (c) ions

10. _____ rays are simply a stream of electrons.

   (a) Alpha   (b) Beta   (c) Gamma

11. _____ rays have the least penetrating power.

   (a) Alpha   (b) Beta   (c) Gamma

12. The _____ is the quantum of light.

   (a) electron   (b) photon   (c) quark

(continued on next page)

Directions: Decide whether each statement that follows is true (**T**) or false (**F**). Write the correct letter in each blank.

_____ 13. It takes less than a year for light from other stars to reach Earth.

_____ 14. The effects of special relativity are most significant when objects move close to the speed of light.

_____ 15. Astronomers can observe many types of sound waves from distant stars.

_____ 16. The weak force binds quarks together to make protons and neutrons.

_____ 17. No object can ever travel faster than the speed of light.

_____ 18. The speed of light in a vacuum is always the same.

_____ 19. Radioactive tracing uses radioactive isotopes to follow the path of a substance through the body.